Two Manuals Combined into One Book

Combat Risk Management

And

Military Training for Urban Operations

Department of Defense

Pentagon
Publishing

FM 100-14

Field Manual
No. 100- 14

Headquarters
Department of the Army
Washington, DC, 23 April 1998

Risk Management

Contents

DISTRIBUTION RESTRICTION: Approved for public release; distribution is unlimited.

Preface

FM 100-14 applies across the wide range of Army operations. It explains the principles, procedures, and responsibilities to successfully apply the *risk management process* to conserve combat power and resources. The manual applies to both Army and civilian personnel during all Army activities, including joint, multinational, and interagency environments.

The manual is intended to help commanders,[1] their staffs, leaders,[2] and managers develop a framework to make risk management a routine part of planning, preparing, and executing operational missions and everyday tasks. This framework will allow soldiers to operate with maximum initiative, flexibility, and adaptability. Although the manual's prime focus is the operational Army, the principles of risk management apply to all Army activities.

Army operations—especially combat operations—are demanding and complex. They are inherently dangerous, including tough, realistic training. Managing risks related to such operations requires educated judgment and professional competence. The risk management process allows individuals to make informed, conscious decisions to accept risks at acceptable levels.

This manual is not a substitute for thought. Simply reading it will not make one adept in building protection around a mission.[3] Soldiers should compare the doctrine herein against their own experience and think about why, when, and how it applies to their situation and area of responsibility. If the doctrine herein is to be useful, it must become second nature.

The proponent of this manual is HQ TRADOC. Send comments and recommendations on DA Form 2028 directly to Commander, US Army Training and Doctrine Command, ATTN: ATBO-SO, Fort Monroe, VA 23651-5000.

Unless this publication states otherwise, masculine nouns and pronouns do not refer exclusively to men.

---

1 The term *commander* as used herein refers to personnel in a command position.

2 The term *leader* as used herein refers to commanders, personnel in the chain of command (team, squad, section, platoon leader), and staff members having personnel supervisory responsibility.

3 The term *mission* as used herein includes mission, operation, or task.

Introduction

Risk management is not an add-on feature to the decision-making process but rather a fully integrated element of planning and executing operations... Risk management helps us preserve combat power and retain the flexibility for bold and decisive action. Proper risk management is a combat multiplier that we can ill afford to squander.

General Dennis J. Reimer
Chief of Staff, Army
27 July 1995

The Army's fundamental purpose is to fight and win the nation's wars. For this purpose, the country gives the Army critical resources, including those most valuable—its sons and daughters. The Army uses its resources to generate overwhelming combat power to fight and win quickly, decisively, and with minimal losses. The Army's inherent responsibility to the nation is to protect and preserve its resources—a responsibility that resides at all levels. Risk management is an effective process for preserving resources. It is not an event. It is both an art and a science. Soldiers use it to identify tactical and accident risks, which they reduce by avoiding, controlling, or eliminating hazards.

The Army introduced the risk management process into training, the operational environments, and materiel acquisition in the late 1980s. Risk management was originally perceived as solely a safety officer function. However, by the early 1990s, the Army established a goal to integrate risk management into all Army processes and activities and into every individual's behavior, both on and off duty. Since the process was introduced, the personal involvement of commanders in preventing accidents—and their aggressive use of the process—have become driving factors in the steady downward trend in Army accidental losses.

Leaders must understand the importance of the process in conserving combat power and resources. Risk management, like reconnaissance and security, is an ongoing process that continues from mission to mission. Within the mission, leaders must know when the process begins and who has responsibility. It must be integral to the military decision. The process is an important means to enhance situational awareness.

Risk decisions are commanders' business. Such decisions are normally based on the next higher commander's guidance on how much risk he is willing to accept and delegate for the mission. Risk decisions should be made at the lowest possible level, except in extreme circumstances. Training operations, including those at combat training centers (CTCs), may be of such intensity that risk decision are retained at a higher level.

Both leaders and staffs manage risk. Staff members continuously look for hazards associated with their areas of expertise. They then recommend controls to reduce risks. Hazards and the resulting risks may vary as circumstances change and experience is gained. Leaders and individual soldiers become the assessors for ever-changing hazards such as those associated with environment (weather; visibility; contaminated air, water, and soil), equipment readiness, individual and unit experience, and fatigue. Leaders should advise the chain of command on risks and risk reduction measures.

Index

I

K

L

M

N

O

P

R

S

Chapter 1

Risk Management Fundamentals

Sizing up opponents to determine victory, assessing dangers and distances is the proper course of action for military leaders.

Sun Tzu, The Art of War, "Terrain"

Risk management is the process of identifying, assessing, and controlling risks arising from operational factors and making decisions that balance risk costs with mission benefits. Leaders and soldiers at all levels use risk management. It applies to all missions and environments across the wide range of Army operations. Risk management is fundamental in developing confident and competent leaders and units. Proficiency in applying risk management is critical to conserving combat power and resources. Commanders must firmly ground current and future leaders in the critical skills of the five-step risk management process.

Risk is characterized by both the probability and severity of a potential loss that may result from hazards due to the presence of an enemy, an adversary, or some other hazardous condition. Perception of risk varies from person to person. What is risky or dangerous to one person may not be to another. Perception influences leaders' decisions. A publicized event such as a training accident or a relatively minor incident may increase the public's perception of risk for that particular event and time—sometimes to the point of making such risks unacceptable. Failure to effectively manage the risk may make an operation too costly—politically, economically, and in terms of combat power (soldiers lives and equipment). This chapter presents the background, principles, applicability, and constraints relating to the risk management process.

BACKGROUND

Throughout the history of armed conflict, government and military leaders have tried to reckon with the effect of casualties on policy, strategy, and mission accomplishment. Government and military leaders consider battle losses from different perspectives. However, both must balance the following against the value of national objectives:

- Effects of casualties.
- Impact on civilians.
- Damage to the environment.
- Loss of equipment.
- Level of public reaction.

War is inherently complex, dynamic, and fluid. It is characterized by uncertainty, ambiguity, and friction. *Uncertainty* results from unknowns or lack of information. *Ambiguity* is the blurring or fog that makes it difficult to distinguish fact from impression about a situation and the enemy. *Friction* results from change, operational hazards, fatigue, and fears brought on by danger. These characteristics cloud the operating environment; they create risks that affect an army's ability to fight and win. In uncertainty, ambiguity, and friction, both danger and opportunity exist. Hence, a leader's ability to adapt and take risks are key traits. Chapter 2 of FM 100-5 provides information on the challenging circumstances of military operations during conflict.

Historically, the Army has had more accidental losses, including fratricide (friendly fire), than losses from enemy action. See Figure 1-1. These accidental losses are the same types experienced in peacetime

Army	World War II 1942–1945	Korea 1950–1953	Vietnam 1965–1972	Desert Shield/ Storm[1] 1990–1991
Accidents	56%	44%	54%	75%
Friendly Fire	1%	1%	1%	5%
Enemy Action	43%	55%	45%	20%

[1]These numbers include the relatively long buildup time and short period of combat action

Figure 1-1. Battle and Nonbattle Casualties

during training exercises. These losses are not caused by the enemy or an adversary. Factors include—

- An ever-changing operational environment.
- Effects of a fast-paced, high-operational tempo (OPTEMPO) and a high-personnel tempo (PERSTEMPO) on unit and human performance. Examples include leader or soldier error or failure to train or perform to standards.
- Equipment failure, support failure, and the effects of the physical environment.

PRINCIPLES

The basic principles that provide a framework for implementing the risk management process are—

- *Integrating risk management into mission planning, preparation, and execution.* Leaders and staffs continuously identify hazards and assess both accident and tactical risks. They then develop and coordinate control measures. They determine the level of residual risk for accident hazards in order to evaluate courses of action (COAs). They integrate control measures into staff estimates, operation plans (OPLANs), operation orders (OPORDs), and missions. Commanders assess the areas in which they might take tactical risks. They approve control measures that will reduce risks. Leaders ensure that all soldiers understand and properly execute risk controls. They continuously assess variable hazards and implement risk controls.
- *Making risk decisions at the appropriate level in the chain of command.* The commander should address risk guidance in his commander's guidance. He bases his risk guidance on established Army and other appropriate policies and on his higher commander's direction. He then gives guidance on how much risk he is willing to accept and delegate. Subordinates seek the higher commander's approval to accept risks that might imperil the next higher commander's intent.
- *Accepting no unnecessary risk.* Commanders compare and balance risks against mission expectations and accept risks only if the benefits outweigh the potential costs or losses. Commanders alone decide whether to accept the level of residual risk to accomplish the mission.

APPLICABILITY

Risk management applies to all situations and environments across the wide range of Army operations, activities, and processes. Risk management is useful in developing, fielding, and employing the total Army force. Figure 1-2 summarizes the key aspects of risk management.

DEVELOPMENT

Development concerns include force design, manpower allocation, training and training developments, and combat and materiel developments (equipment and weapons systems) and battle laboratories.

Risk management assists the commander or leader in—

- Conserving lives and resources and avoiding unnecessary risk.
- Making an informed decision to implement a COA.
- Identifying feasible and effective control measures where specific standards do not exist.
- Providing reasonable alternatives for mission accomplishment.

Risk management does not—

- Inhibit the commander's and leader's flexibility and initiative.
- Remove risk altogether, or support a zero defects mindset.
- Require a GO/NO-GO decision.
- Sanction or justify violating the law.
- Remove the necessity for standard drills, tactics, techniques, and procedures.

Figure 1-2. Key Aspects of Risk Management

Force Design

Concerns include risks introduced in trade-off decisions that involve the design and equipping of—

- Tables of organization and equipment (TOE).
- Modification tables of organization and equipment (MTOE).
- Tables of distribution and allowances (TDA) organizations.

Manpower Allocations

Concerns include shortfalls in manning that put unit readiness and full use of combat system capabilities at risk.

Training and Training Developments

Concerns include hazardous and critical training tasks and feasible risk reduction measures that provide leaders with the flexibility to safely conduct tough, realistic training.

Combat and Materiel Developments and Battle Laboratories

Concerns include providing a means to assist in making informed trade-off decisions such as—

- Balancing equipment form, fit, and function.
- Balancing the durability and cost of equipment and spare parts against their reliability, availability, and maintainability requirements.
- Determining the environmental impact.
- Determining whether to accept systems with less than the full capabilities prescribed in requirement documents and experimental procedures.

ARs 70-1 and 385-16 and MIL-STD-882 provide details on risk management application in the Army materiel acquisition process.

FIELDING

Fielding concerns include personnel assignments, sustainment and logistics, training, and base operations.

Personnel Assignments

Concerns include making informed decisions in assigning replacement personnel. For example, a risk is associated with assigning a multiple launch rocket system crewmember as a replacement for a tube artillery cannon crewmember.

Sustainment and Logistics

Concerns include enhancing one's ability to determine support requirements, the order in which they should be received, and the potential impact of logistics decisions on operations.

Training

Concerns include helping leaders determine the—

- Balance between training realism and unnecessary risks in training.
- Impact of training operations on the environment.
- Level of proficiency and experience of soldiers and leaders.

Base Operations

Concerns include prioritizing the execution of base operations functions to get the most benefit from available resources. Examples include allocating resources for pollution prevention, correcting safety and health hazards, and correcting violations of environmental protection regulations. FM 20-400 provides specific guidance on environmental protection in military operations.

EMPLOYMENT

Employment concerns include force protection and deployment, operations, and redeployment.

Force Protection

Concerns include developing a plan that identifies threats and their associated hazards and balancing resource restraints against the risk.

Deployment, Operations, and Redeployment

Concerns include—

- Analyzing the factors of mission, enemy, terrain, troops, and time available (METT-T) to determine both tactical and accident risks and appropriate risk reduction measures.
- Determining the correct units, equipment composition, and sequence.
- Identifying controls essential to safety and environmental protection.

CONSTRAINTS

Risk management does not convey authority to violate the law-of-land warfare or deliberately disobey local, state, national, or host nation laws. It does not justify ignoring regulatory restrictions and applicable standards. Neither does it justify bypassing risk controls required by law, such as life safety and fire protection codes, physical security, transport and disposal of hazardous material and waste, or storage of classified material. Commanders may not use risk management to alter or bypass legislative intent. However, when restrictions imposed by other agencies adversely affect the mission, planners may negotiate a satisfactory COA if the result conforms to the legislative intent.

Risk management assists the commander in complying with regulatory and legal requirements by—

- Identifying applicable legal standards that affect the mission.
- Identifying alternate COAs or alternate standards that meet the intent of the law.
- Ensuring better use of limited resources through establishing priorities to correct known hazardous conditions that will result in projects with the highest return on investment funded first.

Chapter 2

Risk Management Process

First reckon, then risk

Field Marshal Helmuth von Moltke

This chapter provides the essence of the five-step risk management process. It illustrates the application of each step to military operations through the factors of METT-T.

THE FIVE STEPS: AN OVERVIEW

Risk management is the process of identifying and controlling hazards to conserve combat power and resources. The five steps of risk management are—

- Step 1. Identify hazards.
- Step 2. Assess hazards to determine risks.
- Step 3. Develop controls and make risk decisions.
- Step 4. Implement controls.
- Step 5. Supervise and evaluate.

This five-step process is integrated into the military decision-making process as shown in Figure 2-1.

FM 100-40 provides insight into the context in which the risk management process is applied herein. Areas of particular interest in FM 100-40 include—

- Solving tactical problems (Chapter 1).
- The science and art of tactics (Chapter 1).
- Hasty versus deliberate operations (Chapter 1).
- The plan-prepare-execute cycle (Chapter 1).
- Basic tactical control measures (Chapter 2).
- The factors of METT-T (Chapter 2).

Military Decision-Making Process	Risk Management Steps				
	Step 1 Identify Hazards	Step 2 Assess Hazards	Step 3 Develop Controls and Make Risk Decision	Step 4 Implement Controls	Step 5 Supervise and Evaluate
Mission Receipt	X				
Mission Analysis	X	X			
COA Development	X	X	X		
COA Analysis	X	X	X		
COA Comparison			X		
COA Approval			X		
Orders Production				X	
Rehearsal[1]	X	X	X	X	X
Execution and Assessment[1]	X	X	X	X	X

[1]All boxes are marked to emphasize the continued use of the risk management process throughout the mission

Figure 2-1. Risk Management Steps Correlated with Military Decision-Making Tasks

Risk decisions should be based upon awareness rather than mechanical habit. Leaders should act on a keen appreciation for the essential factors that make each situation unique instead of from conditioned response. Throughout the entire operational continuum, the commander must consider US Government civilians and contract support personnel in his risk management process. Hazards can exist, regardless of enemy or adversary actions, in areas with no direct enemy contact and in areas outside the enemy's or adversary's influence. The two types of risk that exist across the wide range of Army operations are *tactical risks* and *accident risks.*

- *Tactical risk* is risk concerned with hazards that exist because of the presence of either the enemy or an adversary. It applies to all levels of war and across the spectrum of operations.
- *Accident risk* includes all operational risk considerations other than tactical risk. It includes risks to the friendly force. It also includes risks posed to civilians by an operation, as well as an operations impact on the environment. It can include activities associated with hazards concerning friendly personnel, civilians, equipment readiness, and environmental conditions.

STEPS 1 AND 2

Steps 1 and 2 together comprise the risk assessment. In Step 1, individuals identify the hazards that may be encountered in executing a mission. In Step 2, they determine the direct impact of each hazard on the operation. The risk assessment provides for enhanced situational awareness. This awareness builds confidence and allows soldiers and units to take timely, efficient, and effective protective measures.

STEPS 3 THROUGH 5

Steps 3 through 5 are the essential follow-through actions to effectively manage risk. In these steps, leaders balance risk against costs—political, economic, environmental, and to combat power—and take appropriate actions to eliminate unnecessary risk. During execution, as well as during planning and preparation, leaders continuously assess the risk to the overall mission and to those involved in the task. Finally, leaders and individuals evaluate the effectiveness of controls and provide lessons learned so that others may benefit from the experience.

THE FIVE STEPS APPLIED

STEP 1. IDENTIFY HAZARDS

A *hazard* is an actual or potential condition where the following can occur due to exposure to the hazard:

- Injury, illness, or death of personnel.
- Damage to or loss of equipment and property.
- Mission degradation.

Hazards are sources of danger or risks due to enemy or adversary presence and other conditions not due to enemy or adversary capabilities. Hazards are found in all operational environments. Combat operations, stability operations, base support operations, and training present unique hazards for units involved in these kinds of missions. Hazards are identified during the first four steps of the military decision-making process: *mission receipt*, *mission analysis*, *COA development*, and *COA analysis*.

The ability of unit leaders and staffs to identify hazards is key. One reality of today's missions is that the aspect of a hazard can change rapidly. Things of little risk initially can quickly become major threats due to unforeseen natural or man-made events. Leaders should be aware of this possibility. Complacency to the fact that existing controls may not continue to control hazards in rapidly changing situations should be viewed as a hazard in itself.

The factors of METT-T provide a sound framework for identifying hazards when planning, preparing, and executing operations. When applying risk management to METT-T during mission analysis, leaders and staffs should look for hazards that affect both tactical and accident risks. They must identify all hazards that may present significant risks to the mission.

Mission

Leaders first analyze the assigned mission. They look at the type of mission to be accomplished and consider possible subsequent missions. Certain kinds of operations are inherently more dangerous than others. For example, a deliberate frontal attack, because of the associated movement, is more likely to expose a unit to losses than would a defense from prepared positions. Identifying missions that routinely present great risk is imperative. Leaders also look for hazards associated with complexity of the plan such as—

- A scheme of maneuver that is difficult to understand or too complex for accurate communications down to the lowest level.
- The impact of operating under a fragmentary order (FRAGO).

Enemy

Commanders look for enemy capabilities that pose significant hazards to the operation. For example, "What can the enemy do to defeat my operation?" Common shortfalls that can create hazards during operations against an enemy include failure to—

- Assess potential advantages to the enemy provided by the battlefield environment.
- Fully assess the enemy's capabilities.
- Understand enemy collection capabilities and friendly vulnerabilities to those capabilities.
- Accurately determine the enemy's probable COAs.
- Plan and coordinate active ground and aerial reconnaissance activities.
- Disseminate intelligence about the enemy to lower levels.
- Identifying terrorist threats and capabilities.

Intelligence plays a critical part in identifying hazards associated with tactical risk. Intelligence-preparation-of-the-battlefield (IPB) is a dynamic staff process that continually integrates new information and intelligence that ultimately becomes input to the commander's risk assessment process. Intelligence assists in identifying hazards during operations by—

- Identifying the opportunities and constraints the battlefield environment offers to threat and friendly forces.
- Thoroughly portraying threat capabilities and vulnerabilities.
- Collecting information on populations, governments, and infrastructures.

FMs 34-130 and 34-60, respectively, provide detailed information on IPB and on counterintelligence operations and multidiscipline counterintelligence analysis.

Terrain and Weather

In addition to those due to the enemy or adversaries, the most obvious hazards to military operations are due to terrain and weather. Terrain and weather affect the type of hazard encountered. When the enemy uses terrain to his advantage, the risk is clearly tactical. The aspects of terrain and weather may create situations where accident risks predominate. When looking at this from a purely mission perspective, familiarity of the unit with the terrain and its associated environment must be paramount. Basic issues include—

- How long the unit has operated in the environment and climate.
- Whether the terrain has been crossed before.

`Terrain`. The five main military aspects of terrain—*observation and fields of fire, cover and concealment, obstacles, key terrain and decisive terrain, and avenues of approach (OCOKA)*—can be used to identify and assess hazards impacting on friendly forces. Chapter 2 of FM 100-40 has details on OCOKA. The terrain analysis includes both map and on-the-ground reconnaissance to identify how well unit capabilities and mission demands can be accommodated by the terrain.

- *Observation and fields of fire.* Hazards associated with this usually involve when the enemy will be able to engage a friendly unit and when friendly unit weapons capabilities allow it to effectively engage the enemy.
- *Cover and concealment.* Hazards associated with cover and concealment are created by the enemy's ability to place direct or indirect fire on friendly forces.
- *Obstacles.* Hazards associated with obstacles may be accident or tactical. They may be due to natural conditions such as rivers or swamps or man-made such as minefields or built-up areas.
- *Key terrain and decisive terrain.* Hazards are a marked advantage terrain provides to the enemy if he controls such terrain or denies its use to friendly forces.
- *Avenues of approach.* Hazards associated with avenues of approach can affect both tactical and accident risks. Such hazards include conditions where an avenue of approach impedes deployment of friendly combat power or where it supports deployment of enemy combat power.

`Weather`. Weather works hand-in-hand with terrain to create hazards. To identify weather hazards, leaders and soldiers must assess the impact on operating systems. Mistakes include not considering the—

- Adverse effects of heat and cold hazards on the performance of soldiers.
- Effects of climate and weather on maintenance of vehicles and equipment before beginning an operation.
- Hazardous effects of weather on the five military aspects of terrain.

Troops

Leaders analyze the capabilities of available friendly troops. Associated hazards impact both the soldier and unit. Key considerations are level of training, manning levels, the condition and maintenance of vehicles and equipment, morale, availability of supplies and services, and the physical and emotional health of soldiers. Leaders and soldiers must be vigilant to the fact that hazards in these areas can adversely affect a mission, even when all tactical considerations point to success. Mission failure can be caused by—

- *Hazards to the physical and emotional health of soldiers.* Inadequate sanitation facilities, water purification capabilities, medical attention, and evacuation capabilities are key hazards that can arise from incomplete logistical planning. Care of troops requires long-range projections of all classes of supply, with close monitoring of mission changes that could impact availability or depletion of supplies. When beginning an operation immediately upon arriving in theater, hazards include not implementing measures to help soldiers overcome fatigue or acclimatize them to the geographical area and associated climate.
- *Hazards to task organization or units participating in an operation.* Hazards include how long units have worked together under a particular command relationship. During stability operations, task organizations may change often. Hazards include poor communication, unfamiliarity with higher headquarters SOPs, and insufficient combat power to accomplish the mission.
- *Hazards associated with long-term missions.* Long-term missions include nation building, peacekeeping, or insurgency/counterinsurgency operations. Hazards associated with these missions include the turmoil of personnel turnover, lack of continuity of leadership, inexperience, and lack of knowledge of the situation and the unit's operating procedures. An especially insidious hazard is critical-skills atrophy that results from not performing METL-related missions.

Time Available

The hazard is insufficient time to plan, prepare, and execute operations. Planning time is always at a premium. Leaders routinely apply the one-third/two-thirds rule to ensure their subordinate units are given maximum time to plan. Failure to accomplish a mission on time can result in shortages of time for subordinate and adjacent units to accomplish their mission.

Civilians

The commander's legal responsibility is to consider hazards to, and safeguarding of, civilians in his area of operations. *Civilians* include nongovernmental organizations (NGOs), private voluntary organizations (PVOs), US Government civilians, foreign national civilians, the media, and dislocated civilians put at risk by military operations. The commander must consider hazards that can occur across the range of operations, such as—

- *In a wartime environment.* The commander must consider the hazard of collateral damage which may result in creating new adversaries.
- *In a peacetime environment.* The commander must consider the political attitudes and previous actions of civilians in identifying hazards to friendly forces and the populace itself.

Adversaries are hostile elements other than the enemy that may be encountered during any operation. They present additional hazards. They may be organized opposition or individuals that challenge authority. They may include such diverse elements as rioters, criminals, rogues, or gangs that might want to harass a peace enforcement mission.

STEP 2. ASSESS HAZARDS

Step 2 completes the risk assessment. Risk is the chance of hazard or bad consequences. This step examines each hazard in terms of probability and severity to determine the risk level of one or more hazardous incidents that can result from exposure to the hazard. This step is conducted during three steps of the military decision-making process—*mission analysis, COA development,* and *COA analysis.* This step is also conducted after controls are developed.

The incident must be credible in that it must have a reasonable expectation of happening. The end result is an estimate of risk from each hazard and an estimate of the overall risk to the mission caused by hazards that cannot be eliminated. Leaders must also assess the risk to civilians posed by the operation. They may need to assess the operations' impact on the environment. This step is conducted in three substeps.

Substep A

Leaders and staffs assess each hazard in relation to the *probability* of a hazardous incident. The probability levels estimated for each hazard may be based on the mission, COAs being developed and analyzed, or frequency of a similar event. Figure 2-2 provides a summary of the five degrees of probability. The letters in parentheses following each degree (A through E) provide a symbol for depicting probability. For example, the letter *A* represents *frequent* probability.

FREQUENT (A) Occurs very often, continuously experienced	
Single item	Occurs very often in service life. Expected to occur several times over duration of a specific mission or operation. Always occurs.
Fleet or inventory of items	Occurs continuously during a specific mission or operation, or over a service life.
Individual soldier	Occurs very often in career. Expected to occur several times during mission or operation. Always occurs.
All soldiers exposed	Occurs continuously during a specific mission or operation.
LIKELY (B) Occurs several times	
Single item	Occurs several times in service life. Expected to occur during a specific mission or operation.
Fleet or inventory of items	Occurs at a high rate, but experienced intermittently (regular intervals, generally often,).
Individual soldier	Occurs several times in career. Expected to occur during a specific mission or operation.
All soldiers exposed	Occurs at a high rate, but experienced intermittently.
OCCASIONAL (C) Occurs sporadically	
Single item	Occurs some time in service life. May occur about as often as not during a specific mission or operation.
Fleet or inventory of items	Occurs several times in service life.
Individual soldier	Occurs some time in career. May occur during a specific mission or operation, but not often.
All soldiers exposed	Occurs sporadically (irregularly, sparsely, or sometimes).

Figure 2-2. Hazard Probability

SELDOM (D) Remotely possible; could occur at some time	
Single item	Occurs in service life, but only remotely possible. Not expected to occur during a specific mission or operation.
Fleet or inventory of items	Occurs as isolated incidents. Possible to occur some time in service life, but rarely. Usually does not occur.
Individual soldier	Occurs as isolated incident during a career. Remotely possible, but not expected to occur during a specific mission or operation.
All soldiers exposed	Occurs rarely within exposed population as isolated incidents.
UNLIKELY (E) Can assume will not occur, but not impossible	
Single item	Occurrence not impossible, but can assume will almost never occur in service life. Can assume will not occur during a specific mission or operation.
Fleet or inventory of items	Occurs very rarely (almost never or improbable). Incidents may occur over service life.
Individual soldier	Occurrence not impossible, but may assume will not occur in career or during a specific mission or operation.
All soldiers exposed	Occurs very rarely, but not impossible.

Figure 2-2. Hazard Probability (continued)

Substep B

Substep B addresses the *severity* of each hazard. It is expressed in terms of—

- Degree of injury or illness.
- Loss of or damage to equipment or property.
- Environmental damage.
- Other mission-impairing factors such as lost combat power.

The degree of severity estimated for each hazard may be based on knowledge of the results of similar past events. Figure 2-3 provides a summary of the four degrees of hazard severity. The Roman numerals in parentheses following each degree (I through IV) provide a convenient symbol for depicting severity. For example, *I* represents the *catastrophic* degree of severity.

CATASTROPHIC (I)	Loss of ability to accomplish the mission or mission failure. Death or permanent total disability (accident risk). Loss of major or mission-critical system or equipment. Major property (facility) damage. Severe environmental damage. Mission-critical security failure. Unacceptable collateral damage.
CRITICAL (II)	Significantly (severely) degraded mission capability or unit readiness. Permanent partial disability, temporary total disability exceeding 3 months time (accident risk). Extensive (major) damage to equipment or systems. Significant damage to property or the environment. Security failure. Significant collateral damage.
MARGINAL (III)	Degraded mission capability or unit readiness. Minor damage to equipment or systems, property, or the environment. Lost day due to injury or illness not exceeding 3 months (accident risk). Minor damage to property or the environment.
NEGLIGIBLE (IV)	Little or no adverse impact on mission capability. First aid or minor medical treatment (accident risk). Slight equipment or system damage, but fully functional and serviceable. Little or no property or environmental damage.

Figure 2-3. Hazard Severity

Substep C

In this substep leaders and staffs expand what they understand about probable hazardous incidents into estimates of levels of risk for each identified hazard and an estimate of the overall risk for the operation. Estimating risk follows from examining the outcomes of Substeps A and B; that is, both the probability and severity of hazardous incidents. This substep is more art than science. Much depends on the use of historical lessons learned, intuitive analysis, experience, and judgment. Uncertainty can arise in the assessment of both the probability and severity of a hazardous incident. Uncertainty results from unknowns about a situation; from incomplete, inaccurate, undependable, or contradictory information; and from unforeseen circumstances. Therefore, assessment of risk requires good judgment.

Figure 2-4 is a standardized matrix that can be used to assist in this process. Leaders and staffs enter the estimated degree of severity and probability for each hazard in Substeps A and B from the severity row and probability column, respectively. The point where the severity row and probability column intersect defines the level of risk. For example, if the hazard is estimated to have a *critical* severity (II) and a *likely* probability (B), the level of risk is high (H).

Figure 2-5 provides a summary of the levels of risk. It also provides examples of hazardous incidents for each risk level. Several examples illustrate the trade-off between tactical and accident risks.

Risk Assessment Matrix						
		Probability				
Severity		Frequent A	Likely B	Occasional C	Seldom D	Unlikely E
Catastrophic	I	E	E	H	H	M
Critical	II	E	H	H	M	L
Marginal	III	H	M	M	L	L
Negligible	IV	M	L	L	L	L

E – Extremely High Risk
H – High Risk
M – Moderate Risk
L – Low Risk

Figure 2-4. Risk Assessment Matrix

E - Extremely High: Loss of ability to accomplish the mission if hazards occur during mission. A *frequent* or likely probability of catastrophic loss (IA or IB) or *frequent* probability of *critical* loss (IIA) exists.

Example: A commander finds that one of his implied tasks to attack an objective involves crossing a normally shallow riverbed. After looking at the factors of METT-T, he discovers that three days of intense rain have raised the water level to rise above flood stage, with currents far in excess of his ability to safely ford with armored vehicles. After discussing COAs with his staff, he determines the accident risk is extremely high because of the likely probability and catastrophic severity of losing vehicles and killing soldiers. His conclusions are based on his experience with and knowledge of fording armored vehicles under the existing conditions of water depth and current speed.

H - High: Significant degradation of mission capabilities in terms of the required mission standard, inability to accomplish all parts of the mission, or inability to complete the mission to standard if hazards occur during the mission. *Occasional* to *seldom* probability of catastrophic loss (IC or ID) exists. A *likely* to *occasional* probability exists of a critical loss (IIB or IIC) occurring. *Frequent* probability of *marginal* losses (IIIA) exists.

Example: During a preplanned ambush, the leader discovers that the force he intends to ambush has significantly more combat power than his own force can accommodate. He realizes that he could only delay rather than destroy the enemy. He knows his casualty estimates would be very high if the enemy reorganized and counterattacked. He also knows that the size of the enemy force could seriously impact adjacent units conducting a movement to contact. He determines the situation is *high risk* because he estimates (based on his training and experience) there is a likely probability of the enemy reorganizing and counterattacking and the severity of loss to his unit would be critical.

M - Moderate: Expected degraded mission capabilities in terms of the required mission standard will have a reduced mission capability if hazards occur during mission. An *unlikely* probability of catastrophic loss (IE) exists. The probability of a *critical* loss is *seldom* (IID). *Marginal* losses occur with a *likely* or *occasional* probability (IIIB or IIIC). A *frequent* probability of negligible (IVA) losses exists.

Example: A commander in a defensive position receives a warning order to be prepared to counterattack if the enemy attacks again. He chooses to use pre-positioned ammunition caches to support his defense, as opposed to moving his ammunition resupply forward by truck. He determines that the severity of not having an immediate resupply of ammunition available during the counterattack will have a *critical* impact on his combat power. He realizes that if the enemy forces him to abandon his forward positions, the severity of the loss of his

Figure 2-5. Levels of Risk

ammunition caches will critically impact his combat power. He considers that his unit is deployed in excellent defensive positions. He has repelled two attacks that resulted in the destruction of an estimated 50 percent of the enemy's combat power. He receives information that the probability of the enemy attacking is *likely*, but that the probability of the enemy being reinforced and attacking in overwhelming force is remote *(seldom)*. The commander concludes that the risk of conducting a counterattack with limited ammunition is greater than the *moderate* risk of the enemy pushing him back.

L - Low: Expected losses have little or no impact on accomplishing the mission. The probability of *critical* loss is *unlikely* (IIE), while that of *marginal* loss is *seldom* (IIID) or *unlikely* (IIIE). The probability of a *negligible* loss is *likely* or *less* (IVB through (IVE).

Example: A mechanized task force (TF) conducting a movement to contact in a desert environment is overtaken by nightfall before reaching its limit of advance (LOA). The terrain along the axis of advance is flat and open. Visibility is about 800 meters under a clear sky illuminated by a full moon. Estimates put the enemy, which has been hastily withdrawing for the past three days, at approximately 30 percent strength. Contact has been light with no defensible terrain along the TF's axis. The TF commander considers all the factors. In addition, the TF is 100 percent operational in using night vision devices. The TF commander estimates that it is *unlikely* that his unit will incur losses of *critical* severity by being surprised by the enemy or lose *critical* combat power due to an accident. He estimates the risk to his force in continuing a nighttime movement is *low*.

Figure 2-5. Levels of Risk (continued)

STEP 3. DEVELOP CONTROLS AND MAKE RISK DECISIONS

Risk management is the recognition that decision making occurs under conditions of uncertainty. Decisions must remain consistent with the commander's stated intent and offer a good expectation of success. The risk-taking skill requires competency as a prerequisite.

FM 100-7, Decisive Force:
The Army in Theater Operations, May 1995

Step 3 is accomplished in two substeps: develop controls and make risk decisions. This is done during the COA development, COA analysis, COA comparison, and COA approval of the military decision-making process.

Substep A - Develop Controls

After assessing each hazard, leaders develop one or more controls that either eliminate the hazard or reduce the risk (probability and/or severity) of a hazardous incident. When developing controls, they consider the reason for the hazard not just the hazard itself.

Types of Controls Controls can take many forms, but fall into three basic categories—*educational controls, physical controls,* and *avoidance.*

- *Educational controls.* These controls are based on the knowledge and skills of the units and individuals. Effective control is implemented through individual and collective training that ensures performance to standard.
- *Physical controls.* These controls may take the form of barriers and guards or signs to warn individuals and units that a hazard exists. Additionally, special controller or oversight personnel responsible for locating specific hazards fall into this category.
- *Avoidance.* These controls are applied when leaders take positive action to prevent contact with an identified hazard.

Criteria for Controls To be effective, each control developed must meet the following criteria:

- *Suitability.* It must remove the hazard or mitigate (reduce) the residual risk to an acceptable level.
- *Feasibility.* The unit must have the capability to implement the control.
- *Acceptability.* The benefit gained by implementing the control must justify the cost in resources and time. The assessment of acceptability is largely subjective. Figure 2-6 gives criteria for determining acceptability of controls for each identified hazard.

Support	Availability of adequate personnel, equipment, supplies, and facilities necessary to implement a suitable controls.
Standards	Guidance and procedures for implementing a control are clear, practical, and specific.
Training	Knowledge and skills are adequate to implement a control.
Leadership	Leaders are competent to implement a control.
Individual	Individual soldiers are sufficiently self-disciplined to implement a control.

Figure 2-6. Criteria for Determining Acceptability of Controls

Examples of Controls Examples of controls include—

- Engineering or designing to eliminate or control hazards.
- Selecting a COA that avoids identified hazards.
- Limiting the number of people and the amount of time they are exposed to hazards, consistent with mission requirements.
- Selecting personnel with appropriate mental, emotional, and physical capabilities.
- Providing protective clothing, equipment, and safety and security devices.
- Providing such services as adequate sanitation facilities and water purification capabilities.
- Providing warning signs and signals.
- Scheduling vehicle and aircraft silhouette drills.
- Planning training, including rehearsals, rock drills, battle drills, and so forth.
- Programming communications links for key civilian organizations.
- Establishing battlefield controls such as areas of operations and boundaries, direct fire control measures, fire support coordination measures, rules of engagement, airspace control measures, bridge classification, traffic control, and so forth.
- Developing terrorist attack warning systems and response plans.

The key is to specify who, what, where, when, and how each control is to be used. For example—

- Planning and scheduling intensive threat and friendly vehicle identification refresher training for all antiarmor and air defense weapons crews before the mission reduces the probability of engaging a friendly vehicle or aircraft (fratricide).
- Programming installation of crashworthy passenger seats in the UH-60 Blackhawk, when mission circumstances do not indicate their removal, can reduce the severity of injuries in crashes.
- Requiring soldiers to wear flak vests and helmets during movement to contact, or when riding in vehicles in areas where enemy fire is likely, can reduce the probability and severity of a wound from small arms fire or fragments.

- Establishing strong continuity documents and planning overlap tours for key leaders facilitate smooth transitions during extended operations.

Residual RiskOnce the responsible leader develops and accepts controls, he determines the residual risk associated with each hazard and the overall residual risk for the mission.

- *Residual risk* is the risk remaining after controls have been selected for the hazard. Residual risk is valid (true) only if the controls for it are implemented. As controls for hazards are identified and selected, the hazards are reassessed as in Step 2 and the level of risk is then revised. This process is repeated until the level of residual risk is acceptable to the commander or leader or cannot be further reduced. See Figures A-3 through A-5.
- *Overall residual risk* of a mission must be determined when more than one hazard is identified. The residual risk for each of these hazards may have a different level, depending on the assessed probability and severity of the hazardous incident. Overall residual mission risk should be determined based on the incident having the greatest residual risk. Determining overall mission risk by averaging the risks of all hazards is not valid. If one hazard has high risk, the overall residual risk of the mission is high, no matter how many moderate or low risk hazards are present.

Substep B - Make Risk Decision

A key element of the risk decision is determining if the risk is justified. The commander must compare and balance the risk against mission expectations. He alone decides if controls are sufficient and acceptable and whether to accept the resulting residual risk. If he determines the risk level is too high, he directs the development of additional controls or alternate controls, or he modifies, changes, or rejects the COA.

Leaders can use the risk assessment matrix in Figure 2-4—in conjunction with their commanders' guidance—to communicate how much risk they are willing to delegate. For example, a commander may place constraints on his subordinates that restrict their freedom of action to accept risk in instances where the risk might imperil his intent, his higher commander's intent, or a critical capability of the unit.

STEP 4. IMPLEMENT CONTROLS

Leaders and staffs ensure that controls are integrated into SOPs, written and verbal orders, mission briefings, and staff estimates. The critical check for this step, with oversight, is to ensure that controls are converted into clear, simple execution orders understood at all levels. Implementing controls includes coordination and communication with—

- Appropriate superior, adjacent, and subordinate units and those executing the mission.
- Logistics Civil Augmentation Program (LOGCAP) organizations and civilian agencies that are part of the force.

The media, NGOs, and PVOs must be included in coordination when their presence impacts or is impacted by the force.

Leaders must explain how supervisors will implement controls. Examples of control implementation include—

- Conducting vehicle and aircraft silhouette drills.
- Conducting rehearsals, rock drills, battle drills, and so forth.
- Conducting intensive threat and friendly vehicle identification refresher training for all antiarmor and air defense weapons crews.
- Conducting orientation for replacement personnel.
- Installing and maintaining communications links for key civilian organizations.
- Operating in convoys of four vehicles minimum.
- Carrying weapons and wearing flak jackets and helmets when outside secure compounds.

STEP 5. SUPERVISE AND EVALUATE

Leaders must supervise the execution of their orders. The more untrained the troops, the more detailed this supervision must be.

Infantry in Battle, 1939

During mission preparation and execution, leaders must ensure that their subordinates understand how to execute risk controls. Leaders continuously assess risks during the conduct of operations, especially during long-term missions. Leaders maintain situational awareness. They guard against complacency to ensure that risk control standards are not relaxed or violated. To gain insight into areas needing improvement, leaders must continuously evaluate their units' effectiveness in managing mission risks.

Supervise

Leaders supervise mission rehearsal and execution to ensure standards and controls are enforced. Techniques may include spot-checks, inspections, situation reports and brief-backs, buddy checks, and close supervision. During the mission, leaders continuously monitor controls to ensure they remain effective. They modify them as necessary. Leaders and individuals anticipate, identify, and assess new hazards to implement controls. They continually assess variable hazards such as fatigue, equipment serviceability, and the environment. Leaders modify controls to keep risks at an acceptable level.

During sustained operations, leaders continue planning to ensure that controls emplaced at the beginning of the mission apply to changes in the operation's current situation and to hazardous conditions. Leaders must maintain an extraordinary degree of discipline. They must avoid complacency, which can result from boredom and overconfidence. Leaders must ensure that soldiers do not relax their vigilance due to performing repetitive tasks—despite changing roles and missions, unit turbulence and turnover, and declining skills. Leaders maintain a close overwatch on controls put in place to reduce risks over a prolonged period. For example, during stability operations, land mine hazards may not be solved in the near term, but may require continual attention. Other examples of long-term hazards that may be encountered include—

- Climatic extremes.
- NBC and hazardous waste contamination.
- Diseases native to a particular area of operation or indigenous population.
- Terrorist threats.

Evaluate

After a mission, leaders and individuals evaluate how well the risk management process was executed. They—

- Determine how to ensure that successes are continued to the next mission.

- Capture and disseminate lessons learned so that others may benefit from the experience.
- Consider the effectiveness of the risk assessment in identifying and accurately assessing the probability and severity of hazards that resulted in mission degradation.
- Determine whether the level of residual risk of each hazard and of the overall mission were accurately estimated.
- Evaluate the effectiveness of each control in reducing or removing risk, including whether controls were effectively communicated, implemented and enforced.

Leaders and individuals determine why some controls were ineffective and what should be done when the hazard is encountered again. A control may be altered; the way it is implemented or supervised may be changed to make it effective; or a completely different control may be more effective. Leaders must energize the system to fix systemic problems that hinder combat effectiveness.

Figure 2-7 shows that the risk management process continues throughout a mission as well as from mission to mission. It is integral to the military decision-making process. Its application requires good judgment and intuitive analysis borne of confidence, experience, and situational awareness.

TOOLS AND PITFALLS

The appendix provides examples of risk management tools to help leaders assess identified hazards, develop controls, and make risk decisions. The tools should be tailored to suit particular situations and missions. The examples in Figures A-3 through A-5 are tools to manage risk at the tactical level. The example in Figure A-6 is a tool to manage risk at the operational level. Units may develop additional tools suitable for their needs.

Units train to a standard. They operate and train regardless of the degree of real or perceived difficulty. Risk reduction begins with commanders identifying their METLs. Commanders use the risk management process to assess the degree of risk related to each METL their unit must perform. From this assessment, risk reducing standard operating procedures evolve.

Figure A-7 provides an example of risk management considerations integrated into a mission training plan (MTP) task.

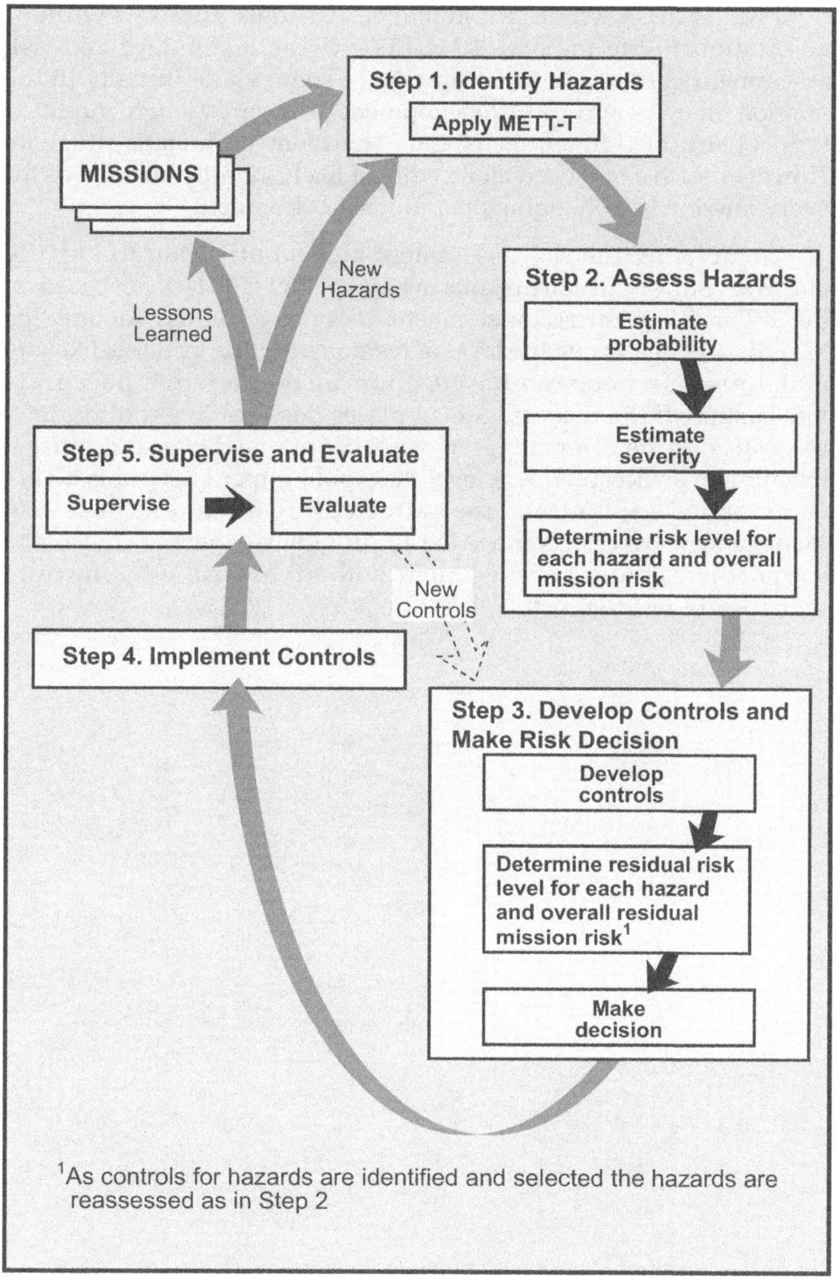

Figure 2-7. Continuous Application of Risk Management

Pitfalls arise when risk management tools are used without adaptation to the factors of METT-T. Using a standardized risk assessment card or checklist may be of some value initially in the mission analysis and COA development or in cases where a routine task is performed in an unchanging environment or static situation. However, such a tool used alone will not likely identify all hazards for every mission in a changing operational environment.

Completing the risk assessment alone, but failing to identify effective controls, usually results in a GO or NO-GO decision based on the initial risk. If the risk assessment does not accurately identify the hazards and determine the level of residual risk, the leader is likely to make his risk decision based upon incomplete or inaccurate information. If the risk assessment places missions in a routine, low-risk category, the commander may not be informed of a risk decision resulting in an accepted risk level that could imperil his or his higher commander's intent or other affected organizations. The risk management process is intended to provide reasonable controls to support mission accomplishment without exposing the force to unnecessary residual risk.

Chapter 3

Risk Management Implementation

It is imperative to develop twenty-first century leaders who trust their subordinates' abilities and judgment. Such leaders must be willing to underwrite their subordinates' honest errors and coach them on to excellence, without tolerating incompetence or laxity. We must recognize that Army leaders are not perfect, and that activity at the ragged edge of audacity sometimes leaves debris in its wake.

GEN William W. Hartzog
Commanding General
US Army Training and Doctrine Command

This chapter presents the moral and ethical implications of risk management. It outlines essential responsibilities and considerations necessary to effectively integrate and assess risk management within the Army. Leaders should tailor these essentials to develop specific how-to procedures suited to their circumstances, available resources, and mission.

MORAL AND ETHICAL IMPLICATIONS FOR LEADERS

To be successful, risk management must be underwritten by the chain of command. Leaders should not expect that all missions will be accomplished with zero defects—free from errors, flaws, or less-than-perfect performance. Demanding such rigid standards leads to oversupervision and paralysis; it produces timid leaders, afraid to make tough decisions in crisis and unwilling to take risks necessary for success in military operations. A zero defects mindset creates conditions that will lead inevitably, in the larger sense, to failure in battle and higher casualties. Leaders are morally bound to support a subordinate's decision to accept risks that are within his commander's intent and guidance, as he understands it.

Leaders accept that things may go wrong, even with the certain knowledge that a subordinate has done all within his power to prevent an incident. In such an event, the leader steps forward and accepts the responsibility along with his subordinate. Furthermore, risk management does not justify taking actions to facilitate an unethical or immoral action. FM 22-100 addresses the moral and ethical aspects of protecting the force within the overall framework of how to apply leadership at all levels to meet mission requirements.

> *Everyday as we respond to the nation's needs, we expose our soldiers to hazards in uncertain and complex environments. We do this with the full knowledge that there are inherent risks associated with any military operation. The nature of our profession will not allow for either complacency or a cavalier acceptance of risk.*
>
> General Dennis J. Reimer
> Chief of Staff, Army

RESPONSIBILITIES

One critical task for all operations is minimizing risk. Every military plan must make this a priority. It is an inherent part of every mission and a basic responsibility of commanders. Whether for training or operational deployments, commanders issue clear risk guidance. Minimizing risk—eliminating unnecessary risk—is the responsibility of everyone in the chain of command. This responsibility runs from the highest commander, through his subordinate leaders, to the soldier.

The commander and his staff must look at tactical risks and accident risks. They use the same risk management process to manage both types. Commanders—with the assistance of their leaders and staffs—manage accident risks. Commanders determine how and where they are willing to take tactical risks.

Leaders and soldiers at all levels are responsible and accountable for managing risks by ensuring that hazards and associated risks are—

- Identified during planning, preparation, and execution of operations.
- Controlled during preparation and execution of operations.

Soldiers are responsible for executing risk controls to standards. They must continuously assess variable hazards such as fatigue, equipment serviceability, and the environment. They must take care of one another and make risk decisions consistent with the higher commander's guidance.

Sometimes commanders are not properly advised in situations where the assumption of risk may affect or imperil their units, the intent of their higher commander, or the operations of an adjacent unit. This is most often attributed to—

- The *risk denial syndrome* in which leaders do not want to know of the risk.
- A staff member who believes that the risk decision is part of his job and does not want to bother his commander or section leader.
- A subordinate who does not fully understand the higher commander's guidance.
- Complacency—outright failure to recognize a hazard or the level of risk involved, or overconfidence in one's abilities or the unit's capabilities to avoid or recover from a hazardous incident.
- Use of a standardized risk assessment tool, such as a risk assessment card, that is not tailored to the unit's mission or adapted to the factors of METT-T and which may put missions in the routine low-risk category.

COMMANDERS

The commander directs the organization and sets priorities and the command climate (values, attitudes, and beliefs). Successful preservation of combat power requires embedding risk management into unit behavior. This requires commitment and creative leadership—innovative planning, careful management. It also requires the chain of command's demonstrated support of the risk management process. Only then will the Army begin to capture the full power of risk management. Commanders establish a command climate favorable for risk management integration by—

- Demonstrating consistent and sustained risk management behavior through leading by example—habitually doing risk management—and actively participating throughout the risk management process.
- Providing clear guidance, when appropriate, on where or what risk to accept.

- Obtaining and providing to subordinates the necessary assets to control risk.
- Knowing their own limitations, their leaders' and soldiers' limitations, and their unit's capabilities.
- Preventing a zero-defects mindset from creeping into their command's culture.
- Allowing subordinates to make mistakes and learn from them.
- Demonstrating full confidence in subordinates' mastery of their trade and their ability to execute a chosen COA.
- Keeping subordinates informed; consulting with subordinate leaders before making a decision, if feasible.
- Listening to subordinates.

Commanders are responsible and accountable for their own actions and those of units under their charge. Commanders must weigh the repercussions of casualties, damage to the environment, and loss of equipment. They must also consider the level of public reaction to loss against national, strategic, operational, or tactical objectives. Commanders are also responsible for keeping soldiers from falling into complacency. During sustained operations in particular, complacency can creep in as a result of accepting the status quo. For example, a unit can be deployed for several months and nothing may happen. The unit then gets very satisfied with itself and its performance. It gets conditioned to its initial circumstances—being well-armed, well-equipped, well-disciplined, and well-trained. It thinks it has risks under control and does not need to change. Such complacency, and the associated loss of situational awareness, can result in leaders and soldiers taking a gamble instead of a prudent risk. A risk is the accepted result of an informed decision; a gamble is an uninformed bet or guess on a hopeful outcome. Leaders and soldiers must clearly understand the difference.

Command is often exercised in conditions of uncertainty and ambiguity, where violence, danger, fear, and friction abound, and under the ever present time constraints driven by OPTEMPO. Risk decisions are frequently required by and dependent on the immediate situation. Judgment is required; a formula, rule, or checklist, by itself, is not appropriate under such circumstances.

Avoiding the zero-risk mindset requires the exercise of positive leadership. The commander's approach to managing risk should be through empowering leaders by pushing risk decisions as far down the chain of command as feasible within the next higher commander's guidance. Commanders must insist that subordinates exercise their freedom of action to act decisively and aggressively to complete assigned missions and promote success of the larger force. Subordinates must consider hazards outside their assigned responsibilities that impact the mission. The result is to encourage coordination and communication—laterally and up and down the chain of command. This requires and encourages initiative, which demands well-trained, determined, disciplined soldiers.

Risk management is a two-way street. It is important that those involved in mission preparation and execution be fully aware of the amount of command involvement and actions necessary to control or remove hazards. The higher commander's guidance specifies the degree of damage or risk to subordinate units that he is willing to accept during the current operation. Subordinates ensure they understand and implement their commander's intent and guidance. If, during the planning process, the accident and/or tactical residual risk exceeds that which the higher commander is willing to accept, the subordinate informs his commander. He requests the resources necessary to mitigate the risk. If, during mission execution, the subordinate determines the risk is too great, he directs the development of additional or alternate controls or modifies or changes the COA. He should notify his next higher commander of his decision. Requiring subordinates to report to the higher commander when a risk decision point is reached during mission execution can result in paralysis.

The objective of managing risk is not to remove all risk, but to eliminate unnecessary risk. Commanders conduct tough, realistic training, knowing that they may put lives and property at risk in the course of military operations. Nothing is worth the cost of a life as the result of taking unnecessary risk. If an action will result in an unacceptable risk, measures should be taken to mitigate it. If the risk cannot be mitigated to an acceptable level, the action should not be executed. Circumstances may occur during mission execution when a decision to stop and defer execution of the operation should be made to avoid taking unwarranted risk. Such a situation will generally occur at the tactical level. For example, circumstances may determine if a trade-off between maintaining the momentum of the attack or risking fratricide or serious accidents is justified. For example, during the deployment of TF Eagle to Bosnia-Herzegovina, there was pressure to complete the Sava River bridge as quickly as possible. The commander assessed the situation and decided to call for a short rest period because his soldiers were tired and getting careless. Consequently, they completed the bridge ahead of schedule under the most difficult conditions imaginable and without injuries.

LEADERS

Many of the tasks identified for commanders apply to all leaders. Leaders' specific responsibilities in managing risk include—

- Establishing clear, feasible risk management policies and goals.
- Conducting detailed planning within time constraints; assessing each mission and task in terms of its risk; continuously reassessing risk as the mission and conditions change and experience is gained.
- Making informed risk decisions and establishing and clearly communicating risk guidance.
- Training the risk management process. Ensuring subordinates understand the who, what, where, when, how, and why of managing risk and how the process applies to their circumstances and assigned responsibilities.
- Examining how subordinates manage risk and how soldiers protect themselves.
- Supervising and evaluating the unit's execution of risk controls during the mission to correct areas needing improvement.
- Advising his chain of command on risks and risk-reduction measures.
- Providing subordinates with feedback on their performance and ways to improve.
- Assessing the effectiveness of their unit's risk management program.
- Capturing and disseminating lessons learned to ensure they are continued from mission to mission so that others may benefit from the experience.

STAFFS

The chief of staff or executive officer is responsible for supervising integration of risk management across the staff. He coordinates development of risk reduction controls with emphasis on deconflicting controls that affect multiple functional areas and adjacent units. The staff officer helps the commander eliminate unnecessary risks by—

- Analyzing his functional area and applying risk management during the military decision-making process.
- Identifying constraints in the higher commander's risk guidance.
- Including hazards and their risks in the mission analysis briefing.
- Including a risk assessment for the commander's estimate.
- Considering the risk assessment in the operations estimate.
- Including risks and recommending ways to reduce their impact in the staff estimate.
- Implementing risk controls by coordinating and integrating them into the appropriate paragraphs and graphics of the OPORD and into products such as SOPs and OPLANs.
- Establishing procedures and standards that are clear and practical.
- Determining the effectiveness of hazard/risk controls and continuously assessing their suitability, feasibility, and acceptability.
- Supervising, evaluating, and assessing the integration of risk management during an operation.
- Continuously identifying hazards, assessing initial and residual risks for each hazard, recommending control measures to reduce the risk to the force.
- Identifying and assessing hazards associated with complacency, especially during extended operations, and recommending appropriate actions to the commander.

At the operational level, staffs focus on hazards and their risks across the spectrum of protecting the force

INDIVIDUALS

The level of expertise and maturity of the individual influence his proficiency in managing risk. Managing risk is subjective because its basis is individual judgment. Young soldiers are routinely charged with executing hazard controls and risk reduction measures. By nature, some them are impulsive risk takers. Their limited experience, coupled with an infallible attitude, can significantly increase the level of risk they are willing to accept. Their sense of indestructibility, motivation *(esprit de corps),* and willingness to achieve the mission at any cost also play a part.

Some soldiers and individuals ignore existing standards and improvise their own. Due to inexperience or complacency, they become susceptible to—

- Overestimating their ability to respond to or recover from a hazardous incident—they become overconfident.
- Underestimating the level of risk posed by a hazard.

It is imperative that individuals understand and execute controls implemented by leaders and staffs.

Individuals must maintain situational awareness and self-discipline when they perform their duties. They must—

- Understand and apply risk management.
- Execute controls directed by their leaders, that is, *perform to standards.*
- Carry risk management over into training and activities—both on and off duty.
- Look out for others—anyone has authority to halt something that is inherently unsafe.

INTEGRATION INTO TRAINING AND OPERATIONS

ARFOR commanders/leaders must continuously employ risk management approaches to effectively preclude unacceptable risks to personnel and property, including protecting forces preparing for or en route to combat.

FM 100-7, *Decisive Force: The Army in Theater Operations,* May 1995

Integrating risk management into training and operations—

- Preserves the lives and well-being of everyone.
- Conserves equipment, facilities, environmental resources, and combat power.

Risk management must not be treated as an afterthought. It must be planned for up front. Leaders and managers of materiel acquisition, base operations, and industrial operations must budget risk control costs up front at the level of expected payback over the duration of the activity, or the life cycle of materiel/weapons system.

When integrating risk management into sustained operations, leaders must consider increases in turbulence, personnel turnover, critical skill atrophy, and mission development. Leaders must continuously assess—

- The complexity of mission development and associated changing interrelationships with other agencies.
- The inclusion of civilian contractors, for example, LOGCAP, as part of the force.
- The presence of the media, NGOs, and PVOs.

These diverse elements need to be integrated into the risk management process.

Two key considerations relevant to managing risk in complex operational environments include—

- Understanding the culture of the indigenous population or society and its way of doing business. Leaders should respect their way of life and not interfere with local customs. Such interference could risk damage to relationships and increase the potential for introducing instability into the local society. Leaders must not, however, intentionally allow these considerations to endanger their force or its mission.
- The dynamics of managing risk and the way the leader and his subordinates manage risk. Leaders manage risk by—
  - — Having the right combination of well-trained, disciplined, well-armed, and well-equipped forces.
  - — Issuing clear guidance to minimize risk.
  - — Determining and implementing risk controls for carrying out the mission.

— Maintaining situational awareness, especially of vulnerabilities.

— Avoiding unnecessary accident risk.

Risk management works best when incorporated into existing command training and operational planning cycles. They then act as risk management tools and provide for—

- Development of policy, goals, objectives, and priorities in the commander's quarterly training guidance.
- The commander's training assessment, by identifying hazards and risk controls.
- Systematic observation and assessment of the unit's risk management performance and feedback into the training management cycle and SOPs.

Leaders and soldiers must have the skills, knowledge, and attitude to effectively manage risks inherent in all operations. Effective training helps soldiers become proficient. It qualifies them technically and tactically, and as leaders, to accomplish the mission without unnecessary risk.

Doctrine developers must integrate risk management into planning for all Army processes, especially the training management cycle defined in FMs 25-100 and 25-101. Unit leaders and their staffs must continually assess and evaluate the integration of risk management into short-, near-, and long-term training plans. They must continually review METLs to ensure that training is supported by realistic risk management objectives. In the past, unprepared or improperly trained units paid a high price for veteran status. The Army must learn from past experiences if it is to avoid repeating such losses.

ASSESSMENT OF THE RISK MANAGEMENT PROCESS

To assess the risk management process is to determine a unit's current level of proficiency in implementing the process. The term *assessment,* as discussed here, differs from *evaluation* as used in Step 5 of the process. Evaluation is used to measure demonstrated ability to accomplish specified objectives within a discrete operation or exercise. Assessment, as used here, also differs from the aspect of assessing hazards for probability and severity, as addressed in Step 2 of the process.

How well risk is managed affects readiness. Leaders need to know the current status and effectiveness of their organization's risk management program. They self-assess their unit's effectiveness in managing risk in order to gain insight into areas for improvement and get feedback on subordinates' understanding and application of risk guidance. The assessment objectives are to determine how—

- Effectively risk management is embedded into planning and preparing for operations.
- Well risk management is understood by subordinate leaders and soldiers.
- Effectively risk management is used to execute operations.

Leaders assess the effectiveness of their units by reviewing how well hazards are identified and risk controls are—

- Specified in oral and written OPORDs, OPLANs, and SOPs.
- Communicated to lowest level of chain of command.
- Included in short-, near-, and long-term training plans.
- Implemented into training and activities on and off duty.
- Embedded into protect-the-force programs such as safety and health and antiterrorism.
- Part of after-action reviews and fed into lessons learned.

Risk management cannot be seen as a competitive program whereby a unit or leader is judged or compared in a competitive sense. Focus is strictly on both reduction of risk and risk behavior.

Take calculated risks. That is quite different from being rash.

General George S. Patton, Jr.

Examples of Risk Management Application

The examples in this appendix are designed to help those charged with managing risk.

TRACKING TOOL

The work sheet instructions are in Figure A-1.

Work Sheet Instructions	
Block	
A – D	Self explanatory
E	Identify task relating to the mission or task in Block A
F	**Identify Hazards** – Identify hazards by reviewing METT-T factors for the mission or task. Additional factors include historical lessons learned, experience, judgment, equipment characteristics and warnings, and environmental considerations.
G	**Assess Hazards** – Assessment includes historical lessons learned, intuitive analyses, experience, judgment, equipment characteristics and warnings, and environmental considerations. Determine initial risk for each hazard by applying risk assessment matrix (Figure 2-4). Enter the risk level for each hazard.
H	**Develop Controls** – Develop one or more controls for each hazard that will either eliminate the hazard or reduce the risk (probability and/or severity) of a hazardous incident. Specify who, what, where, why, when, and how for each control. Enter controls.
I	**Determine Residual Risk** – Determine the residual risk for each hazard by applying the risk assessment matrix (Figure 2-4). Enter the residual risk level for each hazard.
J	**Implement Controls** – Decide how each control will be put into effect or communicated to the personnel who will make it happen (written or verbal instruction; tactical, safety, garrison SOPs, rehearsals). Enter controls.
K	**Determine Overall Mission/Task Risk** – Select the highest residual risk level and circle it. This becomes the overall mission or task risk level. The commander decides whether the controls are sufficient to accept the level of residual risk. If the risk is too great to continue the mission or task, the commander directs development of additional controls or modifies, changes, or rejects the COA.
	Supervise and Evaluate – This last step is not on the worksheet. Plan how each control will be monitored for implementation (continuous supervision, spot-checks) and reassess hazards as the situation changes. Determine if the controls worked and if they can be improved. Pass on lessons learned.

Figure A-1. Risk Management Work Sheet Instructions

The work sheet (Figure A-2) provides a starting point to logically track the process of hazards and risks. It can be used to document risk management steps taken during planning, preparation, and execution of training and combat missions and tasks.

A. Mission or Task:		**B. Date/Time Group** **Begin:** **End:**		**C. Date Prepared:**	
D. Prepared By: (Rank, Last Name, Duty Position)					
E. Task	**F. Identify Hazards**	**G. Assess Hazards**	**H. Develop Controls**	**I. Determine Residual Risk**	**J. Implement Controls** ("How To")
K. Determine overall mission/task risk level after controls are implemented (circle one) LOW (L) MODERATE (M) HIGH (H) EXTREMELY HIGH (E)					

Figure A-2. Sample Risk Management Work Sheet

Examples provided in Figures A-3 through A-6 should help individuals manage risk at the tactical level.

A. Mission or Task: Prepare defensive positions	**B. Date/Time Group** **Begin:** 010035R May XX **End:** 010600R May XX	**C. Date Prepared:** 29 April XX
D. Prepared By: (Rank, Last Name, Duty Position) LT Jones, Plt Ldr		

E. Task	**F. Identify Hazards**	**G. Assess Hazards**	**H. Develop Controls**	**I. Determine Residual Risk**	**J. Implement Controls** ("How To")
Construct nonstandard antivehicular wire obstacle	Back injuries and wire cuts during materiel offload	Moderate (M)	Use proper lift and carry methods and wear concertina wire gloves and safety goggles	Low (L)	Unit TACSOP, ARTEP 5-145 DRILL (pg 2-44) ARTEP 5-335-11-MTP
	Blunt trauma and cuts in pounding of U-shaped pickets	Moderate (M)	Wear helmet and increase situational awareness	Low (L)	Unit TACSOP, ARTEP 5-145 DRILL (pg 2-44) ARTEP 5-335-11-MTP
	Cuts when unrolling concertina	Moderate (M)	Wear concertina wire gloves and maintain situational awareness	Low (L)	Unit TACSOP, ARTEP 5-145 DRILL (pg 2-44) ARTEP 5-335-11-MTP
	Cuts when installing concertina	Moderate (M)	Wear concertina wire gloves and maintain situational awareness	Low (L)	Unit TACSOP, ARTEP 5-145 DRILL (pg 2-44) ARTEP 5-335-11-MTP
	Cuts when installing barbed wire	Moderate (M)	Wear concertina wire gloves and maintain situational awareness	Low (L)	Unit TACSOP, ARTEP 5-145 DRILL (pg 2-44) ARTEP 5-335-11-MTP

K. Determine overall mission/task risk level after controls are implemented (circle one)

LOW (L) (circled) MODERATE (M) HIGH (H) EXTREMELY HIGH (E)

Figure A-3. Example of Completed Risk Management Work Sheet for Squad/Platoon

E. Task	F. Identify Hazards	G. Assess Hazards	H. Develop Controls	I. Determine Residual Risk	J. Implement Controls ("How To")
A. Mission or Task: Conduct a deliberate attack			**B. Date/Time Group** Begin: 010035R May XX End: 010600R May XX		**C. Date Prepared:** 29 April XX
D. Prepared By: (Rank, Last Name, Duty Position) CPT William Wallace, Cdr					
Conduct obstacle breaching operations	Obstacles	High (H)	Develop and use obstacle reduction plan	Low (L)	Unit TACSOP, OPORD, training handbook
	Inexperienced soldiers	High (H)	Additional instruction and increased supervision	Moderate (M)	Modified training schedule, additional instruction
	Operating under limited visibility	Moderate (M)	Use NVDs; use IR markers on vehicles	Low (L)	Unit TACSOP, OPORD
	Steep cliffs	High (H)	Rehearse use of climbing ropes	Moderate (M)	FM 90-6, *Mountain Operations* TC 90-6-1, *Mountaineering*
	Insufficient planning time	High (H)	Plan and prepare concurrently	Moderate (M)	OPORD, troop-leading procedures
K. Determine overall mission/task risk level after controls are implemented (circle one) LOW (L) MODERATE (M) [circled] HIGH (H) EXTREMELY HIGH (E)					

Figure A-4. Example of Completed Risk Management Work Sheet for Company/Team

A. Mission or Task: Redeploy unit to home station			B. Date/Time Group Begin: 010035R May XX End: 010600R May XX	C. Date Prepared: 29 April XX	
D. Prepared By: (Rank, Last Name, Duty Position) MAJ Woolsey, S3					
E. Task	**F. Identify Hazards**	**G. Assess Hazards**	**H. Develop Controls**	**I. Determine Residual Risk**	**J. Implement Controls** ("How To")
Conduct convoy operations-movement from Grafenwohr to home base	Fatigue leading to poor judgment/ accident-causing errors	Moderate (M)	1. Ensure all drivers receive adequate rest the night before movement 2. Brief planned rest stops and actions to take if driver becomes too tired/ill or has doubts about alertness 3. Have NCOs check physical alertness of drivers prior to movement	Low (L)	Unit SOP, OPORD on Tactical Road March, STP 21-2-MQS (Tactical Road March)
	Fast moving traffic mix of cars and large trucks with trailers	Moderate (M)	1. Use extreme caution when passing 2. Continue to scan (be aware of) traffic 3. Brief all drivers and vehicle commanders on lane-changing procedures	Low (L)	German laws and unit SOP
	Accidents and breakdowns blocking road	Moderate (M)	1. Brief drivers on actions to take a. Pull as far off the road as possible b. Get out on passenger side c. Post triangle sign at least 100 meters behind vehicle d. Stay as far away from vehicle and traffic as possible e. Brief all personnel on maintenance plan/action if vehicle is disabled f. Brief all personnel on procedures for vehicle exits on the traffic side	Low (L)	Unit SOP, OPORD on Tactical Road March

Figure A-5. Example of Completed Risk Management Work Sheet for Battalion/Task Force

A. Mission or Task: Redeploy unit to home station (continued)			B. Date/Time Group Begin: 010035R Dec XX End: 010600R Dec XX	C. Date Prepared: 29 Nov XX	
D. Prepared By: (Rank, Last Name, Duty Position) MAJ Woolsey, S3					
E. Task	**F. Identify Hazards**	**G. Assess Hazards**	**H. Develop Controls**	**I. Determine Residual Risk**	**J. Implement Controls** ("How To")
Conduct convoy operations-movement from Grafenwohr to home base	Accidents and breakdowns blocking road (continued)	Moderate (M)	2. Leaders ensure that proper maintenance are done prior to move (PMCS), dispatching and licensing procedure. 3. Contact team follows convoy and repairs or removes disabled vehicles from highway.	Low (L)	
	Weather conditions– high winds, sleet, snow	High (H)	1. Reduce speed according to environmental conditions less stopping distance available. 2. Road conditions especially right after rains – hydroplaning – review procedures for skids or hydroplanes. 3. Maintain convoy interval to allow for braking and assist civilian traffic in passing.	Moderate (M)	Unit SOP and vehicle operations manual
	High winds– reduction visibility, sudden reduction due to passing traffic	Moderate (M)	1. Change lanes carefully; ensure clearance and avoid erratic moves 2. Be aware of winds and effects on vehicles. 3. Drive defensively.	Low (L)	German traffic laws
	20% downgrade at checkpoint 1	Moderate (M)	1. Use lower gears. 2. Reduce speed to 25 mph	Low (L)	Unit SOP and vehicle operations manual

Figure A-5. Example of Completed Risk Management Work Sheet for Battalion/Task Force (continued)

A. Mission or Task:		B. Date/Time Group		C. Date Prepared:	
Redeploy unit to home station (continued)		**Begin:** 010035R Dec XX **End:** 010600R Dec XX		29 Nov XX	
D. Prepared By: (Rank, Last Name, Duty Position) MAJ Woolsey, S3					
E. Task	**F. Identify Hazards**	**G. Assess Hazards**	**H. Develop Controls**	**I. Determine Residual Risk**	**J. Implement Controls ("How To")**
Conduct convoy operations-movement from Grafenwohr to home base	Heavy traffic, traffic jams, and congestion	Moderate (M)	1. Use extreme caution when passing 2. Use headlights at all times 3. Brief convoy speeds and vehicle intervals 4. Brief actions to take if vehicle pulls in front a. Maintain clearance from vehicle b. Maintain braking distance 5. Stay in right lane in construction zones 6. Ensure AMVs 2 or more meters wide remain in right lane 7. Continuously scan surroundings (situational awareness)	Low (L)	Provide risk assessment to convoy leaders. Provide safety brief information to leaders and drivers
K. Determine overall mission/task risk level after controls are implemented (circle one) LOW (L) MODERATE (M) (circled) HIGH (H) EXTREMELY HIGH (E)					

Figure A-5. Example of Completed Risk Management Work Sheet for Battalion/Task Force (continued)

A. Mission or Task: Peace Enforcement			B. Date/Time Group Begin: 010035R May XX End:		C. Date Prepared: 29 April XX
D. Prepared By: (Rank, Last Name, Duty Position) G-3 Div/Corps/EAC					
E. Task	**F. Identify Hazards**	**G. Assess Hazards**	**H. Develop Controls**	**I. Determine Residual Risk**	**J. Implement Controls** ("How To")
Occupy area of operations	Assault on facilities	Moderate (M)	Identify and isolate combatants, terrorists, supporters Harden potential terrorist targets	Low (L)	Coordination with local law enforcement agencies FM 5-103, *Survivability* *General /Order*
	Ambush	Moderate (M)	Kevlar helmets and flak jackets required outside cantonment area Minimum of 4 vehicles for movement Two qualified drivers per vehicle	Low (L)	Rehearsals Immediate action drills General orders
	Mines	High (H)	Countermine plan Mine awareness training Predeployment training	Moderate (M)	FM 20-32, *Mine/Countermine Operations* Lessons learned

Figure A-6. Example of Completed Risk Management Work Sheet for Division/Corps/EAC Task Force

A. Mission or Task: Peace Enforcement			B. Date/Time Group Begin: 010035R May XX End:		C. Date Prepared: 29 April XX
D. Prepared By: (Rank, Last Name, Duty Position) G-3 Div/Corps/EAC					
E. Task	**F. Identify Hazards**	**G. Assess Hazards**	**H. Develop Controls**	**I. Determine Residual Risk**	**J. Implement Controls** ("How To")
Occupy area of operations	Fratricide	Moderate (M)	Identify friend/foe Antifratricide training LNO augmentations	Low (L)	Lessons learned ROE Weapons control procedures AOR orientation
	Seasonal risk (cold weather injuries)	Moderate (M)	Cold weather clothing and equipment Soldier/leader awareness training	Low (L)	Severe weather alert/action plan TB Med 507
	Seasonal risk (hot weather injuries)	Moderate (M)	Hot weather clothing and equipment Soldier/leader awareness training	Low (L)	Severe weather alert/action plan TB Med 507
	Disease	Moderate (M)	Immunizations Field sanitation Preventive medicine	Low (L)	Spring thaw plan Focused medical training Vector control program

Figure A-6. Example of Completed Risk Management Work Sheet for Division/Corps/EAC Task Force (continued)

A. Mission or Task: Peace Enforcement		B. Date/Time Group Begin: 010035R May XX End:		C. Date Prepared: 29 April XX	
D. Prepared By: (Rank, Last Name, Duty Position) G-3 Div/Corps/EAC					
E. Task	**F. Identify Hazards**	**G. Assess Hazards**	**H. Develop Controls**	**I. Determine Residual Risk**	**J. Implement Controls** ("How To")
Occupy area of operations	Vehicle accidents	Moderate (M)	Driver qualifications Driver fatigue reduction Convoy safety procedures Monitor route conditions Vehicle inspections	Low (L)	Driver qualification training Driver sleep plans Two qualified drivers per vehicle Convoy safety briefings Route reconnaissance Vehicle inspections
K. Determine overall mission/task risk level after controls are implemented (circle one) LOW (L) MODERATE (M) [circled] HIGH (H) EXTREMELY HIGH (E)					

Figure A-6. Example of Completed Risk Management Work Sheet for Division/Corps/EAC Task Force (continued)

MISSION TRAINING PLAN

Figure A-7 shows risk management integrated into an MTP task. The example portrays a possible method for integrating the risk management process and products into MTPs. The text in bold italics indicates suggested areas in which to integrate risk management.

TASK: C^2 THE BATTALION (7-1-1901)(FM 7-20)

ITERATION: 1 2 3 4 5 (circle)

TRAINING STATUS: T P U (circle)

CONDITION: The brigade issues an OPORD or FRAGO

TASK STANDARD :

a. The battalion plan accomplishes the directed mission and specified tasks according to the brigade commander's concept and intent. The plan is received and understood by the leadership of the battalion, who makes the plan successful. It is coordinated with higher, adjacent, and supporting elements.

b. The plan is as fully prepared as time allows to initiate the mission at the directed time.

c. The battalion controls and synchronizes subordinate and supporting elements so that it accomplishes the mission and preserves the force. ***Include commander's risk guidance.***

d. The battalion keeps higher, adjacent, subordinate, supporting, and supported headquarters informed of essential information key to controlling the battle or making required decisions.

SUBTASKS AND STANDARDS: GO NO-GO

+1. Battalion leaders issue the warning order.

a. A complete warning order is issued within 15 minutes of receipt of the brigade order.

b. Warning order is received by all platoons within 45 minutes of issuance of battalion warning order.

+ Critical task.
* Leader task.

Figure A-7. Integration of Risk Management into MTP Task

***2. Battalion commander analyzes mission and gives initial guidance.**

a. Guidance includes restated mission, which includes brigade commander's intent for battalion and identifies all specified and implied tasks.

b. Guidance is given within 30 minutes of receipt of order.

c. Guidance includes instructions on information requirements and initially required preparation actions (movement, resupply) to start. ***Guidance includes chain of command authorized to accept risk (extremely high, high, moderate, and low) affecting higher command's intent, resources, or adjacent units' missions.***

+3. Battalion accomplishes reconnaissance and other actions to gather needed information.

a. Reconnaissance actions begin to physically gain information on the enemy and terrain as early as possible.

b. Commander conducts a personal reconnaissance, when possible. If not, the commander conducts a detailed map reconnaissance.

c. Subordinate leaders perform a personal reconnaissance when possible. See subordinate company (ARTEP 7-10-MTP) and platoon (ARTEP 7-8-MTP) T&EOs.

d. Staff coordinates with subordinate, higher, supporting, and adjacent headquarters to gather information for planning. ***Staff analyzes mission to identify hazards.***

e. Staff provides operations, intelligence, and CSS estimates to include all critical METT-T factors and ***risk assessment considerations (identified hazards/risks).***

+4. Battalion commander develops and wargames courses of action and selects one.

Figure A-7. Integration of Risk Management into MTP Task (continued)

a. Tactically feasible courses of action (include CS and maneuver) are made and wargamed with the available staff (commander, S3, and FSO are best for quick planning sequences; XO, S2, engineer, S4, S3 (Safety), and ADA officer are best in more deliberate situations). ***Each course of action contains assessment of hazards, risk level, and control measures identified to lower or control the risk.***

b. Best COA is selected.

c. COA is wargamed and refined by the command and staff. The staff must understand the concept to produce a sound OPORD and rehearse. ***The staff continually assesses the situation to identify new hazards, assess the risk from each hazard, develop acceptable control measures, and reassess for residual risk to the mission. COA should pose minimum risk to soldiers, equipment, and mission accomplishment. Risk acceptance decisions for the COA are made at the appropriate level in the chain of command.***

*5. Staff develops an OPLAN and OPORD from the commander's guidance.

OPLAN and OPORD successfully accomplish the mission according to higher commander's intent. ***Hazards and risk control measures are included in the appropriate paragraphs and graphics.***

*6. Battalion commander and staff issue implementing FRAGOs (accident risk controls).

*7. Battalion commander issues the OPORD/FRAGO.

a. OPORD/FRAGO is issued IAW the one-third, two-thirds rule and makes full use of daylight time.

b. OPORD/FRAGO accomplishes all directed missions and tasks, complies with the brigade commander's intent, controls risk, and is doctrinally sound. It is based on evaluator judgment and on comparison of brigade and battalion OPORDs.

Figure A-7. Integration of Risk Management into MTP Task (continued)

c. All subordinate and supporting elements receive the OPORD/FRAGO.

d. OPORD/FRAGO contains task organization, mission, concept, ***accident risk controls,*** intent for maneuver, supporting fires, and obstacles; missions and tasks for each subordinate; fire support and CSS instructions; and coordinating instructions to synchronize the efforts of maneuver forces and CS.

e. If more time is available, the battalion commander issues a fully developed OPORD (although an initial FRAGO may be issued to allow subordinates to begin preparation and followed by a full OPORD.)

f. Order is given at a location that reduces travel time, allows observation of the zone/sector, and promotes OPSEC. (Depending upon the METT-T factors, observation of the zone/sector may not be possible.)

g. Battalion commander should perform brief-backs and war gaming, ***to include identifying hazards and control measures,*** immediately after the order to improve subordinate understanding and reaction.

h. Subordinate leaders and staff should perform lateral coordination before leaving the orders site.

+8. Commander and staff coordinate and refine the plan.

a. Time is well used to continue gathering information and to improve the plan (contingency plans, ***hazard identification and controls,*** fire plans) .

b. New information is disseminated and coordinated with higher, adjacent, and supporting headquarters to include—

- Changes or refinements in plan.
- Information on the enemy in the sector or zone.

Figure A-7. Integration of Risk Management into MTP Task (continued)

- Information that impacts on planning and execution *(risk acceptance decisions/hazard controls)* of subordinate elements.
- Adjustments/changes in the plans.

+*9. Battalion executes changes in task organization.

a. Main CP coordinates link-up location, time, and responsible element.

b. Attachments/new elements are received at the coordinated location and time; updated on current situation, OPORDs, and SOIs; and resupplied.

c. Detachments reach the link-up point at the time and place directed.

+*10. Battalion performs and commander and staff perform, supervise, and monitor preparations.

a. Command group/XO performs brief-backs with subordinate commanders, leaders, and key staff.

b. Main CP maintains status of preparations.

c. Elements make full use of time to prepare for the operation. Subjective judgment of the evaluator is based on the analysis of preparation charts and available preparation time.

+*11. Battalion sees the battlefield.

a. Command group is positioned to see and move.

b. Companies and other subordinates accurately report critical information on actions and changes in combat status within five minutes. See subordinate MTPs.

c. Main CP collects, analyzes, and passes processed critical information.

d. Subordinates execute intelligence collection plan. See

Figure A-7. Integration of Risk Management into MTP Task (continued)

subordinate element MTPs. ***Subordinates integrate the risk management process when developing plans and executing tasks.***

+*12. Battalion leaders command and control the execution.

a. Subordinate elements report enemy and friendly actions, change in status, and any other factor that would require change within three minutes.

b. Battalion leaders win the battle by directing the maneuver of units, controlling direct and indirect fires, ***properly integrating risk management into planning, preparation, and execution,*** and directing other CS actions to cope with new METT-T/risk factors. Indicators are:

(1) Elements not following OPLAN and OPORD are corrected.

(2) Responses to new METT-T hazards are directed soon after the new situation occurs.

(3) No friendly casualties inflicted by friendly direct or indirect fires or other accidents.

(4) Number/percentage of direct fire weapons engaging the enemy.

(5) Number/percentage of indirect fire rounds fired and percentage hitting/suppressing the enemy.

(6) Number of enemy casualties.

(7) Number of friendly casualties.

c. The C^2 and CSS assets are controlled to support maneuver effort. Indicators are—

(1) Effective CSS and C^2.

(2) C^2 or CSS elements not destroyed by enemy direct fires.

d. FRAGOs are clear, concise, ***include risk management,*** and are quickly executed by subordinates.

Figure A-7. Integration of Risk Management into MTP Task (continued)

e. Changes that affect the battle are disseminated within five minutes.

+13. Subordinate commanders, leaders, and staff laterally coordinate actions during the battle.

All battle actions requiring coordination between elements are coordinated.

+*14. Battalion coordinates with adjacent and supporting headquarters.

All battle actions requiring coordination with other headquarters are laterally and promptly coordinated.

+*15. Battalion reports.

Battalion CPs submit all critical and required reports to brigade. They report events to adjacent and supporting elements that impact on them in time for those units to react. ***They advise the chain of command, as appropriate, on risks and risk-reduction measures.***

Figure A-7. Integration of Risk Management into MTP Task (continued)

Glossary

AAR after-action review

ADA air defense artillery

ARFOR army forces headquarters

ARTEP Army Training and Evaluation Program

assessment an analytical process to determine an organization's current levels of proficiency on a specific objective (for example, a training objective or risk management implementation) (CJCSM 3500.03)

base operations support the provision of administrative and logistical services; includes supply operations, maintenance of materiel, personnel support, base services and administrative services rendered by or through activities of the supporting installation

BASOPS base operations support

C^2 command and control

COA course of action

CofS chief of staff

combat power the total means of destructive and/or disruptive force that a military unit or formation can apply against an opponent at a given time; a combination of the effects of maneuver, firepower, protection, and leadership

controls actions taken to eliminate hazards or reduce their risk

CP command post

CSS combat service support

CTC combat training center

danger	exposure or vulnerability to harm or risk; the balance between the chance or probability of a hazardous incident and the result of the hazardous incident
EAC	echelons above corps
evaluation	the process used to measure the demonstrated ability to accomplish specified objectives such as training within a discrete event or exercise (CJCSM 3500.03); measurement of the demonstrated ability of soldiers or units to perform a task and supporting skill and knowledge or learning objective against the established standard
exposure	the frequency and length of time personnel and equipment are subjected to a hazard
FM	field manual
FRAGO	fragmentary order
fratricide	the employment of friendly weapons and munitions with the intent to kill the enemy or destroy his equipment or facilities, which results in unforeseen and unintentional death or injury to friendly personnel
friction	the accumulation of chance errors, unexpected difficulties, enemy actions, and confusion of battle
FSO	fire support officer
G3	general staff operations section
hazard	any actual or potential condition that can cause injury, illness, or death of personnel, damage to or loss of equipment, property or mission degradation (FM 101-5); a condition or activity with potential to cause damage, loss or mission degradation (Joint Pub 1-02)
inherently dangerous	an activity or task containing a danger to life or limb that is a permanent and inseparable element of the activity
IPB	intelligence-preparation-of-the-battlefield
IR	infrared

LOA	limit of advance
LOGCAP	Logistics Civil Augmentation Program
METL	mission-essential task list
METT-T	mission, enemy, terrain, troops, and time available
MOS	military occupational specialty
MTP	mission training plan
NBC	nuclear, biological, chemical
NVD	night vision device
OCOKA	O - observation and fields of fire, C - cover and concealment, O - obstacles, K - key terrain and decisive terrain, A - avenues of approach
OPCON	operational control
operational tempo	the pace of an operation or operations; OPTEMPO includes all of the activities the unit is conducting; OPTEMPO can be a single activity or a series of operations
OPLAN	operations plan
OPORD	operations order
OPTEMPO	operational tempo
personnel tempo	unit work load level and number of deployed days per year
PERSTEMPO	personnel tempo
PIR	priority intelligence requirements
probability	the likelihood that a hazardous incident will occur
PVO	private voluntary organization
residual risk	the level of risk remaining after controls have been identified and selected for hazards that may result in loss of combat power
risk	chance of hazard or bad consequences; the probability of exposure to chance of injury or loss from a hazard; risk level is expressed in terms of hazard probability and severity (FM 101-5)

risk assessment	identification and assessment of hazards (first two steps of risk management process); an identified hazard is assessed to determine the risk (both the probability of occurrence and resulting severity) of a hazardous incident due to the presence of the hazard
risk decision	the decision to accept or not accept the risks associated with an action; made by the commander, leader, or individual responsible for performing that action
risk management	the process of identifying, assessing, and controlling risks arising from operational factors and making decisions that balance risk cost with mission benefits
risk management integration	the embedding of risk management principles and practices into Army operations, culture, organizations, systems, and individual behavior
S3	operations officer
severity	the expected consequence of an event (hazardous incident) in terms of degree of injury, property damage, or other mission-impairing factors (loss of combat power and so on) that could occur
situational awareness	ability to have accurate and real-time information on friendly, enemy, neutral, and noncombatant locations; a common, relevant picture of the battlefield scaled to specific level of interest and special need
SOP	standing operating procedure
T&EO	test and evaluation outline
TACSOP	tactical standing operating procedure
TC	training circular
TF	task force
US	United States
XO	executive officer

References

SOURCES USED

FM 20-400. *Military Environmental Protection.* 1997 (draft).

FM 22-9. *Soldier Performance in Continuous Operations.* 12 December 1991.

FM 22-100. *Army Leadership.* 1997 (draft).

FM 25-100. *Training the Force.* 15 November 1988.

FM 25-101. *Battle-Focused Training.* 30 September 1990.

FM 34-60. *Counterintelligence.* 5 February 1990.

FM 34-130. *Intelligence Preparation of the Battlefield.* 23 May 1989.

FM 100-5. *Operations.* August 1997 (draft).

FM 100-40. *Tactics.* 29 October 1997 (draft).

FM 100-7. *Decisive Force: The Army in Theater Operations.* 31 May 1995.

FM 101-5. *Staff Organization and Operations.* 31 May 1997.

READINGS RECOMMENDED

Army Regulation 70-1. *Systems Acquisition Policy and Procedures.* 1997.

Army Regulation 385-16. *System Safety Engineering and Management.* 3 May 1990.

CSA Statement on Risk Management. 27 July 1995.

HQDA Letter 5-97-1. Risk Management Integration Responsibilities. 1 May 1997.

MIL-STD-882C. *System Safety Program Requirements.* 19 January 1993.

"Risk Management for Brigades and Battalions." *Center for Army Lessons Learned (CALL) Newsletter 95-9.* June 1995.

*TC 90-1

Training Circular
No. 90-1

Headquarters
Department of the Army
Washington, DC, 19 May 2008

Training for Urban Operations

Contents

Page

Distribution Restriction: Approved for public release; distribution is unlimited.

*This publication supersedes TC 90-1, 1 April 2002.

Figures

Tables

Chapter 1
Introduction

This chapter discusses definitions, training strategy, prerequisite training, individual task training, and collective task training for urban operations (UO). Military operations on urbanized terrain (MOUT) and UO differ in ways that reflect the changes in the way we engage in war today.

Section I. DEFINITIONS

This section defines and compares MOUT and UO.

Note: The terms "urban operations" and "UO" replaces all instances of "military operations on urbanized terrain" and "MOUT."

MILITARY OPERATIONS ON URBANIZED TERRAIN

1-1. In the past, Army doctrine characterized urban areas as a type of specialized terrain, like jungles and mountains, and called operations in this type of terrain *military operations on urbanized terrain* (MOUT). Doctrine emphasized tactical warfighting tasks on this terrain, which was characterized by densely situated buildings and streets, and assumed the enemy consisted of conventional (Soviet-type) forces. Defeating or destroying those forces overrode all other considerations.

URBAN OPERATIONS

1-2. The concept of UO assumes that the urbanized terrain is populated, and that the populace must be a foremost consideration. UO recognizes that the populace can help either side, and that it requires a working infrastructure. Urban operations, whether against an insurgent or unconventional enemy, thus vastly differ from MOUT (Table 1-1).

Table 1-1. Comparison of MOUT and UO.

	MOUT	*UO*
Enemy force	Conventional, Soviet type	Insurgent or unconventional
Mission	Defeat or destroy enemy above all other considerations	Defeat the enemy Gain and retain the good will of the populace
Populace	Not considered	Important consideration

1-3. Although certain techniques and tactics are common to both types of operations, UO goes beyond MOUT, focusing not only on defeating the enemy, but also on gaining and retaining the good will of the population. UO consider the complex and dynamic interactions and relationships between the urban area's key components--the terrain (natural and manmade), the population, and the supporting infrastructure--as an overlapping and interdependent system of systems.

1-4. A single urban operation may include the full range of Army operations--offensive, defensive, stability and support (Appendix C)--that may be executed, either sequentially or simultaneously (FM 3-06, Chapter 1). Furthermore, to accomplish an objective, UO demand working with sister services, US government agencies, nongovernmental organizations, and regional and international organizations.

Section II. TRAINING STRATEGY

The UO training strategy is primarily organized by environment and phase. Training environments include live, virtual, and constructive (Figure 1-1), and are based on the unit's METL. Training phases are initial, intermediate, and culminating. Table 1-2 shows the relationship between the training environments and the first three training phases. (See also Appendix D, which discusses leader training.)

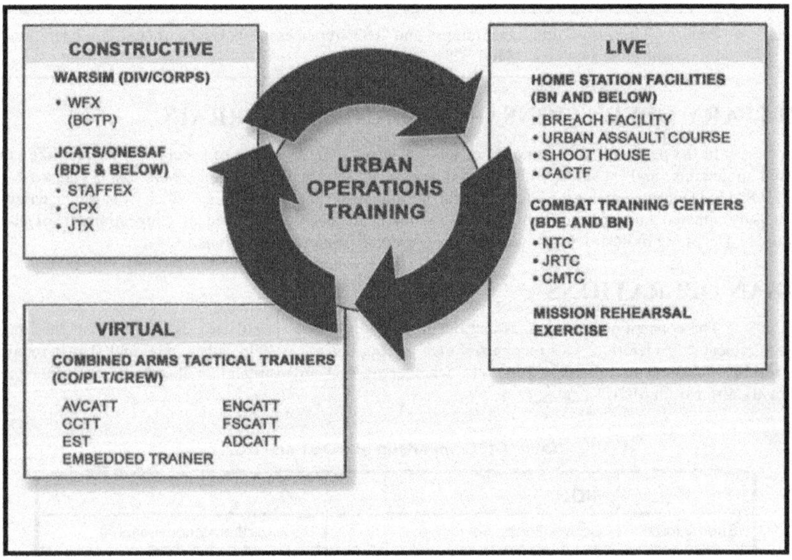

Figure 1-1. Urban Operations Training Support Strategy.

TRAINING PHASES

1-5. The three training phases are initial, intermediate, and culminating. Each of these includes some live, virtual, and constructive training.

INITIAL PHASE

1-6. This phase, which trains individual and collective tasks, is the first step in the building-block approach. It forms the basis of unit or collective training. For Soldiers and leaders, mastery of individual and crew skills is a prerequisite for UO training. The tasks train basic marksmanship skills and other Soldier tasks needed to perform any tactical operation. Included are training in a live environment (precision marksmanship) and virtual environment (Engagement Skills Trainer [EST], crew simulators, and CCTTs).

Table 1-2. Training environment/training phase matrix.

TRAINING PHASE / TRAINING ENVIRONMENT	INITIAL	INTERMED	CULMINATING
LIVE	Precision Marksmanship (Indiv / Sqd)	Breach Facility, Urban Assault Course, Shoot House (Team / Squad / Platoon)	CACTF, Platoon STX, Company STX, Battalion FTX Deployment CTC Rotation Mission Rehearsal Exercise
VIRTUAL	EST (Indiv / Sqd) Crew Simulators and CCTT (Crews Plt Ldrs, Co Cdrs)		
CONSTRUCTIVE			WARSIM (division/corps), which includes warfighting exercise (battle command training program). JCATS/OneSAF (brigade and below), staff exercises (STAFFEXs), command post exercises (CPXs), and joint training exercises (JTX)

INTERMEDIATE PHASE

1-7. This phase of training is characterized by collective task training in varied conditions such as limited visibility; different entry techniques and points; incorporation of combatants, noncombatants, and OPFOR; use of MILES; and live-fire exercises. It is the vital link between individual tasks and mission accomplishment. It should be conducted at, but is not limited to, the *Breach Facility*, *Live-Fire Shoot House*, and *UAC*. (Section III discusses these facilities in detail.) Many types of organizations can use these facilities. Trainers *must* be certified before conducting demolition training IAW local standing operating procedure (SOP) or ACOM policy. AR 385-63 prescribes safety requirements. Except for Aviation, Armor, and other branches that have no practical use for these facilities, this phase must be completed to advance to the culminating phase.

CULMINATING PHASE

1-8. The Combined Arms Collective Training Facility (CACTF, Figure 1-2) is where the culminating phase of urban operations training ends. This phase can include multiechelon, combined arms, and branch specific STX training may be conducted. Urban operations leader training occurs in leader professional development courses (Appendix D). Units have developed internal urban warfare training programs and TRADOC schools have assembled technical mobile training teams (MTTs) for marksmanship, combatives, and breaching techniques.

EDUCATION PHASE

- Urban warfare reading/video list.
- Urban operations equipment.
- Review of Infantry doctrinal publications (FM 3-21.10, FM 3-06.11, for Infantry), and corresponding publications for other branches.
- Small-unit urban movement techniques and considerations.

PLANNING AND EXECUTION PHASE

- Prepare company OPORD or WFF-related products.
- Brief platoon/company OPORD/WFF-related products to commander or staff representative.
- Participate as briefer in urban environment TEWT.

SUSTAINMENT PHASE

- Participate in company/platoon TEWT or other urban operations event.
- Execute/facilitate a company or platoon TEWT.

TRAINING ENVIRONMENTS

1-9. Both live and virtual training environments are used to train UO.

LIVE ENVIRONMENT

1-10. The live training environment consists of home station training using the Urban Assault Course (UAC), Shoot House (SH), Breach Facility (BF), Combined Arms Collective Training Facility (CACTF), Combat Training Center (CTC) rotations, and any other live training that supports the UO training.

Virtual Environment

1-11. The virtual training environment consists of simulators such as Close Combat Tactical Trainer (CCTT) or Engagement Skills Trainer (EST), which are used to train small-unit leaders and crews. The constructive environment consists of battle staff training for battalion and above staff and leaders such as Joint Conflict and Tactical Simulation (JCATS)/OneSAF.

Constructive Environment

1-12. The constructive training environment consists of the following, which all occur in the culminating training phase:

- WARSIM (division/corps), which includes warfighting exercise (battle command training program).
- JCATS/OneSAF (brigade and below), staff exercises (STAFFEXs), command post exercises (CPXs), and joint training exercises (JTX)

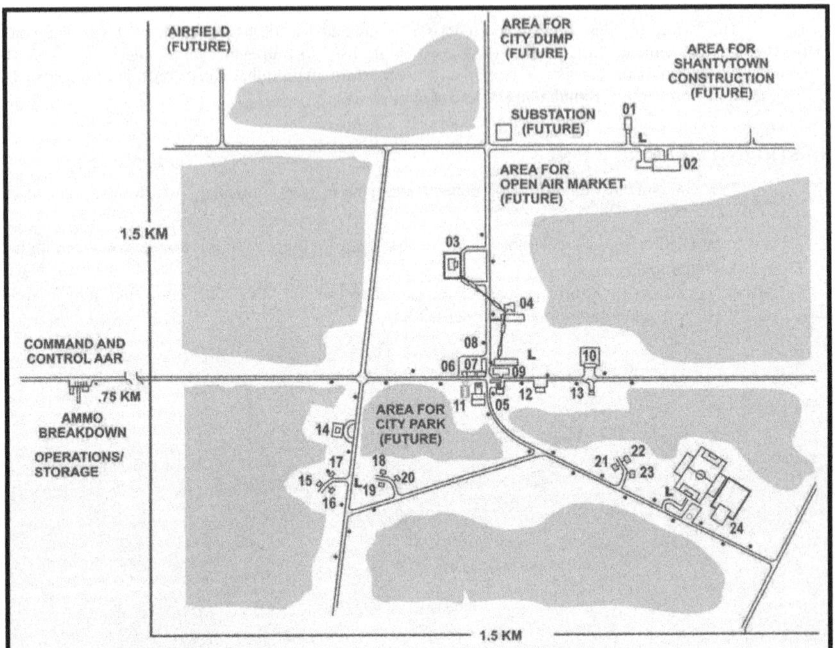

BLDG #	SIZE AND TYPE
01	1 STORY WAREHOUSE
02	1 STORY WAREHOUSE
03	3 STORY OFFICE (EMBASSY)
04	2 STORY OFFICE
05	1 STORY SERVICE STATION
06	1 STORY BUSINESS
07	1 STORY BUSINESS
08	1 STORY BUSINESS
09	3 STORY HOTEL
10	1 STORY POLICE STATION/JAIL
11	CHURCH AND CEMETARY
12	1 STORY BUSINESS
13	1 STORY BANK
14	2 STORY TOWNHOUSE
15	1 STORY RESIDENT
16	1 STORY RESIDENT
17	1 STORY RESIDENT
18	1 STORY RESIDENT
19	1 STORY RESIDENT
20	1 STORY RESIDENT
21	1 STORY RESIDENT
22	1 STORY RESIDENT
23	1 STORY RESIDENT
24	2 STORY SCHOOL

LEGEND

- FENCE WALL
- WALL
- TUNNEL SYSTEM
- TREE LINE
- UTILITY POLE
- LATRINE

SCALE: 1" = 500'

Figure 1-2. Combined Arms Collective Training Facility.

MISSION-ESSENTIAL TASK LIST

1-13. FMs 7-0 and 7-1 explain how to determine a unit's mission-essential task list (METL) and battle tasks, and how to plan, resource, and execute training events. Because training facilities are limited, leaders must use all available time and resources for training and maintaining combat readiness.

- System-specific FMs and TCs teach doctrine, conduct of operations, and employment of weapon systems.
- Tactics, techniques, and procedures (TTP) are lessons learned from training and combat.

1-14. The standards for evaluating training are accessible at the Digital Training Management System (DTMS), https://dtms.army.mil/DTMS.

BATTALION INTEGRATED TRAINING PLAN

1-15. The commander uses the battalion integrated training plan to set goals for training density. He focuses on combined arms operations, emphasizing dismounted Infantry maneuver. Tanks, BFVs, Strykers, attack aviation, and artillery normally provide supporting fires for the dismounted elements. UO training in the CACTF, Shoot House, and Breach Facility should be conducted semiannually and quarterly for the UAC. Army training calendars are divided into two cycles of training: *mission* and *support*. Figure 1-3 and Figure 1-4 show how the urban training strategy and facility usage might be programmed into unit cycles.

RISK ASSESSMENT

1-16. Risk assessment means making operations safe without compromising the level of realistic training (FM 7-1). Commanders continuously assess the risk of training conditions to prevent unnecessary loss of Soldiers and equipment. The degree of risk depends on conditions. Just as this principle applies to other training and warfighting, it also applies to UO. In a training situation, for example, commanders determine whether Soldiers and leaders have conducted this type of training before, whether conducting initial training at night or during limited visibility, and whether Soldiers are tired or exhausted. Risk assessment ensures UO training is not only tough and realistic but also safe. Training enhancers, such as live fire, blank ammunition, simulators, fires, smoke, and other pyrotechnics, must be fully integrated into the training to achieve the training objectives. Soldiers are required to wear earplugs, protective vests, and protective eyewear in order to protect themselves from noise and debris.

1-17. The commander considers risk and safety in urban operations training situations such as--.

- Soldiers engaging in close combat urban operations need to be safety conscious when using blanks, booby traps, and practice grenades. They may be unaware of the exact location of the OPFOR and may be susceptible to OPFOR flash suppressors, booby traps, and practice hand grenades.
- Soldiers should use ladders to enter buildings at upper stories of buildings to prevent unnecessary risks.

1-18. The commander is the safety officer and is responsible for ensuring that his unit, both leaders and Soldiers, comply with safety regulations and the units tactical SOP. All leaders must--.

- Use mission, enemy, terrain, troops, time available, and civilians (METT-TC) factors to identify risks.
- Assess possible losses and their costs.
- Select and develop risk-reduction measures.
- Implement controls by integrating them into plans and orders, SOPs, training performance standards, and rehearsals.
- Supervise and enforce risk reduction measures and safety standards at all times.

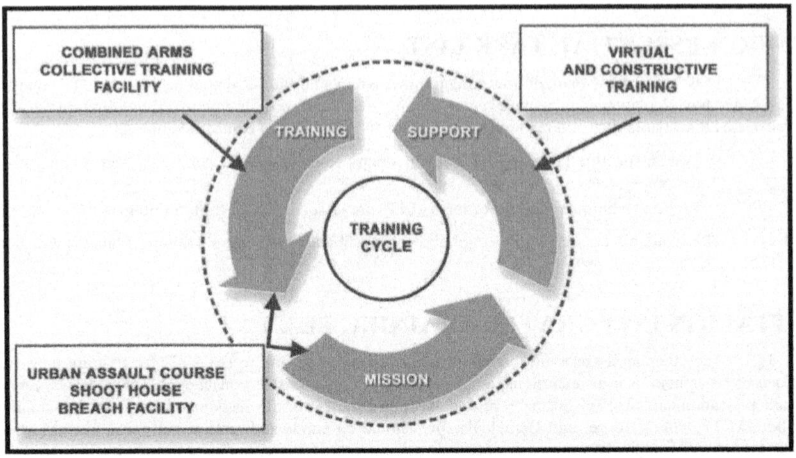

Figure 1-3. Active Army, Home Station, Urban Operations Training Strategy.

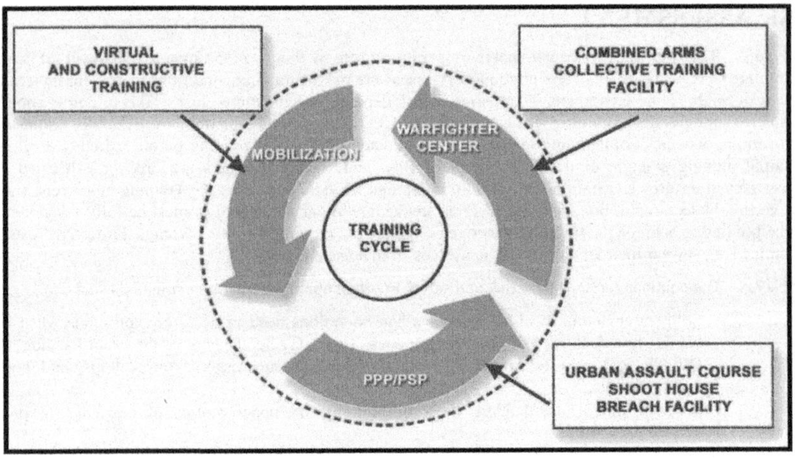

Figure 1-4. Army National Guard Urban Operations Training Strategy.

Section III. TRAINING FACILITIES

Training facilities include three that are discussed in this section: the Urban Assault Course (UAC, Chapter 2), Shoot House (SH, Chapter 3), Breach Facility (BF, Chapter 4). They also include the Combined Arms Collective Training Facility (CACTF, discussed in Chapter 5), Combat Training Centers (CTC), and any other live training that supports the UO training.

URBAN ASSAULT COURSE

1-19. The purpose of the UAC facility is to train a squad-sized force to *Enter a Building and Clear a Room*, improve grenadier gunnery, and validate subterranean TTP. However, the unit's training requirements determine the size of the unit as well as the tasks to be trained. The unit develops its organizational structure and steps for conducting the exercise based on its METL and on their training needs (Figure 1-5). (See Chapter 2 for a more detailed discussion.)

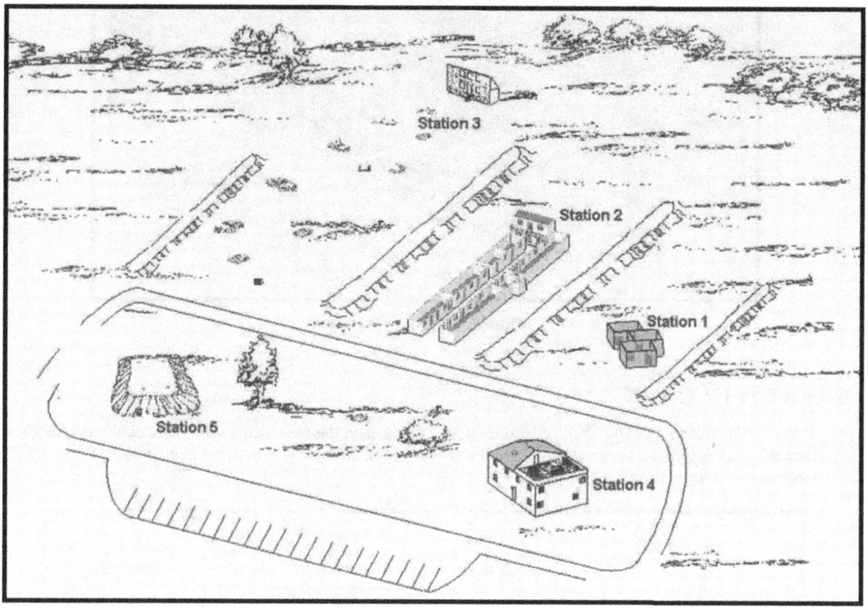

Figure 1-5. Urban Assault Course.

SHOOT HOUSE

1-20. The purpose of this facility is to provide the leader with a facility to train and evaluate fire teams, squads, and platoons during a live-fire exercise (Figure 1-6). Units are trained and evaluated on their ability to conduct tactical movement and engage targets, while practicing target discrimination in an urban environment. (See Chapter 3 for a more detailed discussion.)

Note: The specific designs of the Shoot House and the other facilities may differ from one installation to another, due to variations in training requirements.

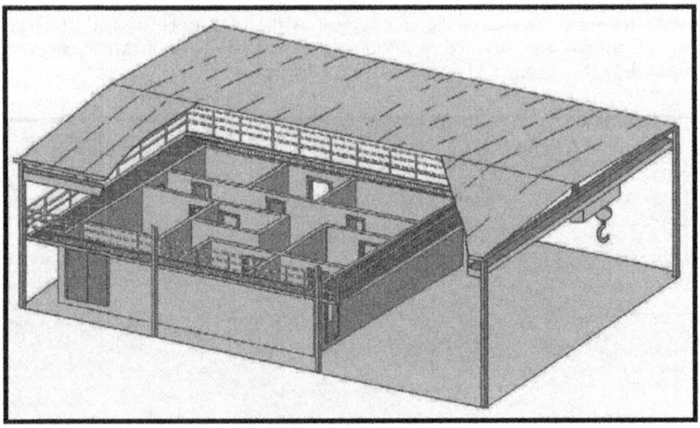

Figure 1-6. Shoot House.

BREACH FACILITY

1-21. The purpose of this facility is to train Warfighters on the technical aspects of mechanical, ballistic, thermal, and explosive breaching techniques (Figure 1-7), including for use in UO. (See Chapter 4 for a more detailed discussion.)

Figure 1-7. Breach Facility.

Section IV. CONDUCT OF TRAINING

Depending on unit proficiency, every training event should follow a *crawl-walk-run* sequence, based on clearly defined task, conditions, and standards. After each training event, leaders conduct an after-action review (AAR).

CRAWL

1-22. The trainer explains each training objective and performance standards, and then talks the Soldiers through the exercise, step by step, describing what each individual or unit must do. This training often includes a demonstration.

WALK

1-23. The Soldiers slowly practice each task to standard. Trainers coach Soldiers at this stage stopping as often as needed to correct mistakes and provide feedback. Soldiers practice each task repeatedly until they can perform it to standard.

RUN

1-24. Soldiers perform each task at full speed as if they are in combat. The Soldiers need feedback during this stage as well. The following can be used to enhance realism:

- Blank ammunition.
- Special Effects Small-Arms Marking System (SESAMS).
- Targetry.
- Pyrotechnics.
- Tactical engagement systems (MILES).
- Opposing forces (force on force).
- Nuclear, biological, and chemical simulation/weapons of mass destruction.
- Training in limited visibility conditions.

AFTER-ACTION REVIEW

1-25. Leaders conduct an after-action review (AAR) after each training event to help Soldiers and units improve their skills by providing immediate feedback. An AAR increases the benefits gained from each training exercise by allowing the leaders and the Soldiers to work together to analyze the performance of each task. Being involved in this analysis enhances the leaders and the Soldiers learning and retention ability. The AAR improves the performance of the leaders and Soldiers in later iterations by allowing interaction and feedback from these individuals during the debrief.

Section V. PREREQUISITE TRAINING

Prerequisite training teaches individuals, crews, units, leaders, and command and staff what they need to know before they are required to perform a task. Prerequisite training builds teamwork and individual confidence at all levels.

COMMAND AND STAFF TRAINING

1-26. The battalion battle staff and leadership must train before the battalion is required to conduct collective training. The battalion command and staff culminates their prerequisite training with a tactical exercise without troops (TEWTs) and command post exercises (CPXs). This sequence of training assists the commander and his staff develop better UO TTP and unit SOP.

LEADER TRAINING

1-27. A comprehensive leader's training program is the key to successful UO training (Appendix D). Time spent teaching and training the battalion leaders (trainers) during the planning and preparation phase will ensure success during the intermediate and culminating phases of training. Realistic training at the unit level will improve the leadership training officers and NCOs receive at the Basic Officer Leadership Course (BOLC), Captains Career Officer Course (CCOC), Basic Noncommissioned Officer Course [BNCOC], and Advanced Noncommissioned Officer Course [ANCOC]. Recommended training should include remedial training on those tasks that the unit had difficulty in during the previous training and evaluation.

INDIVIDUAL TRAINING

1-28. Many individual specific tasks and some collective tasks can be taught and practiced in garrison while most collective tasks are best conducted at the installation UO facilities and weapons ranges. The unit's preliminary individual training tasks should focus on three general areas: urban combat skills, urban marksmanship skills, and physical fitness. The following are example preliminary training tasks or TTP:

- Individual movement techniques.
- Battle drills.
- Weapon positioning in an urban area.
- Roadblock and vehicle search procedures.
- Dismounted and mounted urban navigation.
- Urban scanning, quick-fire, and assault fire techniques.
- Hasty urban firing positions.
- Preparation of urban firing positions.

1-29. Prior to urban collective training cycle, trainers should develop urban-specific firing conditions in conjunction with weapons firing ranges. Build facades on existing ranges, or use the unit's local UAC (if available). Urban combat skills include--

- Firing positions behind rubble, around corners, from windows or rooftops.
- Urban targets behind rubble, in windows and doorways.
- Fire commands.
- Lifting and shifting fires from lower to upper story windows.
- Fire control and fire discipline.
- Rapid firing engagements.
- ROE firing scenarios.

1-30. Urban physical fitness skills include--

- Endurance runs and upper body conditioning.
- Vaulting exercises.

- Urban-specific obstacle course negotiation.
- Leadership Reaction Course negotiation.

INDIVIDUAL TASK TRAINING

1-31. Urban operations require the Soldier to be proficient in those individual tasks that are unique to urban combat. Units should incorporate into their training regime those urban fighting TTP that have proven successful based on lessons learned from combat operations.

TASK FORCE SOLDIER--WARRIOR COMMON TASKS, SKILL LEVELS 1-3:

- 071-326-0541, *Perform Movement Techniques during an Urban Operation.*
- 071-326-0542, *Enter a Building during an Urban Operation.*
- 071-326-0550, *Prepare Positions for Individual and Crew-Served Weapons during an Urban Operation.*
- 071-326-0557, *Select Hasty Firing Positions during an Urban Operation.*
- 071-440-0027, *Execute Mechanical Breach in an Urban Operation.*
- 071-440-0028, *Engage Targets during an Urban Operation.*
- 071-440-0029, *Conduct a Breach during an Urban Operation.*
- 071-440-0003, *Conduct an Attack on a Building by a Squad during an Urban Operation.*
- 071-440-0006, *Conduct a Defense by Squad during an Urban Operation.*

FM 3-06.11, *COMBINED ARMS OPERATIONS IN URBAN TERRAIN*

Movement Techniques

- Crossing of a wall.
- Movement around corners.
- Movement past windows.
- Use of doorways.
- Movement parallel to buildings.
- Crossing open areas.
- Fire team employment.
- Movement between positions.
- Movement inside a building.

Entry Techniques

- Upper building levels.
- Use of ladders.
- Scaling of walls.
- Entry at lower levels.
- Hand grenades.

Firing Positions

- Hasty firing position.
- Prepared firing position.
- Target acquisition.
- Employment of snipers.

COLLECTIVE TASK TRAINING

1-32. Specific FMs teach doctrine, how to conduct operations, and how to employ weapon systems. TTP are lessons learned from the training and combat. The standards for conducting and evaluating collective training are accessible at the Digital Training Management System (DTMS), https://dtms.army.mil/DTMS. Those tasks that do not specifically address UO must be modified for use in an urban environment.

RESOURCES

1-33. Field manuals for the squad and platoon are--:

- FM 3-21.8, *Infantry Rifle Platoon and Squad.*
- FM 3-21.71, *Mechanized Infantry Platoon and Squad (BFV).*
- FM 3-21.9, *SBCT Infantry Rifle Platoon and Squad.*

1-34. Field manuals for the company are--.

- FM 3-21.10, *IBCT Rifle Company.*
- FM 3-21.11, *SBCT Rifle Company.*

1-35. Field manuals and mission training plans for the battalion are--.

- FM 3-21.20, *IBCT Infantry Battalion.*
- FM 3-21.21, *SBCT Infantry Battalion.*
- FM 3-06.11, *Combined Arms Operations in Urban Terrain.*
- TC 7-98-1, *Brigade and Battalion Stability and Support Operations.*

Chapter 2
Urban Assault Course

This chapter describes the Urban Assault Course (UAC), which incorporates doctrine and tactics from FM 3-06.11. The UAC contains five stations:

- Individual and Team Task/Technique Trainer Station 1
- Squad and Platoon Task/Technique Trainer Station 2
- Grenadier Gunnery Trainer Station 3
- Urban Offense/Defense Building Station 4
- Underground Trainer Station 5

These stations are designed for individual tasks and small-unit collective task training. Various types of organizations with a variety of missions can train at the UAC. Tactical operations in an urban environment may include combat to stability and support tasks. (Appendix C provides a SASO unit training matrix). The UAC is a dismounted training facility.

Note: The station discussions are arranged so that they may be removed from the book for ease of use.

PURPOSE

2-1. The purpose is to provide squad and platoon size units with a facility to train and evaluate urban operations tasks (Figure 2-1). Only Station 3, Grenadier Gunnery, is intended for live-fire training. It supports use of 40-mm TP and 5.56-mm service ammunition.

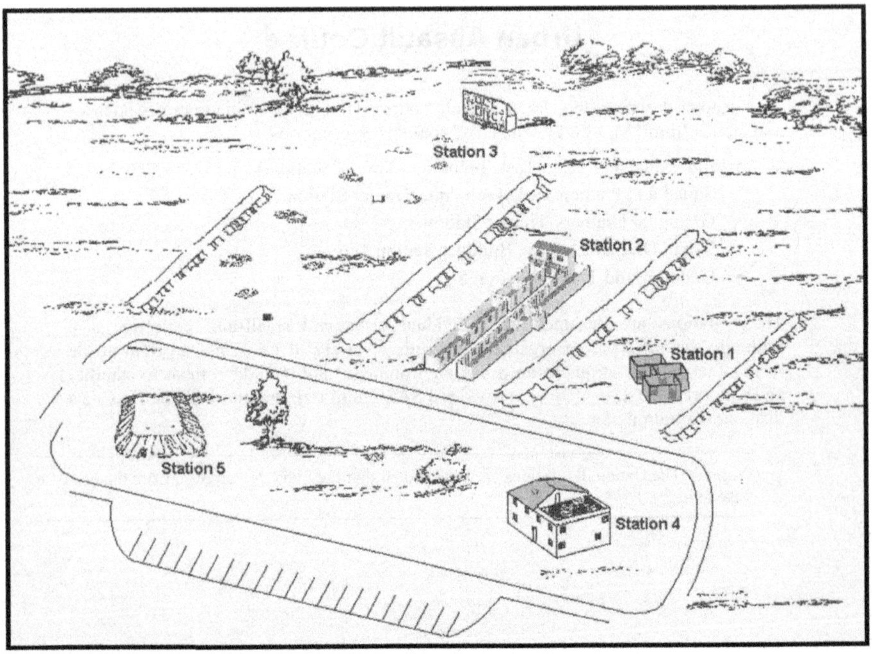

Figure 2-1. Urban Assault Course.

DESCRIPTION

2-2. The UAC has five training stations that facilitate the ***crawl-walk-run*** training concept:

Station 1, Individual and Team Trainer

2-3. This station has three rooms where team leaders and squad leaders train the basics of building and room clearing.

Station 2, Squad and Platoon Trainer

2-4. This station has four structures and multiple rooms. Squads build on tasks learned at Station 1, and begin to learn the concepts of clearing multiple buildings. This station can be used as individual buildings with a narrow street, or as rooms inside a building with a long, connecting hallway.

Station 3, Grenadier Gunnery Trainer

2-5. This is a live-fire station where M203 gunners master target engagements in an urban area, conduct tactical movement, and respond to the fire commands.

Station 4, Offense/Defense House

2-6. This is where a platoon trains to attack or defend a building. This station can also be divided into smaller training stations to reinforce training. It is also used to train tasks not yet trained at the other stations such as upper-level entry techniques.

Station 5, Underground Trainer

2-7. This station provides training for subterranean operations.

TARGETRY

2-8. Targets throughout the UAC are either 3D precision targets (for engagements less than 50 meters) or 2D nonprecision targets (for engagements greater than 50 meters).

Precision Targets

2-9. Precision targets support short distance engagement sensors. These sensors, located inside the precision targets, respond only to lethal shot placements.

- Torso lethal zone .... 18" wide x 8" high
- Neck-head lethal zone .... 4" wide x 8" high

Nonprecision Targets

2-10. Nonprecision targets support engagements at ranges greater than 50 meters. These targets are 2D pop-up type targets.

TRAINING AUDIENCE

2-11. The facility gives Infantry, Engineer, Military Police, Cavalry, and other units a facility for training individual Soldiers through platoon-level UO skills.

TRAINING INFORMATION

2-12. Combat units can use this facility as a mission rehearsal exercise site. Recommended training frequency is quarterly. For each UAC station, they should provide the purpose, a description, targetry, controller and safety requirements, individual UO tasks to be trained, training support requirements, and conduct of training information, if applicable:

Purpose

2-13. This UAC station [*state purpose of station*]. [*Units may use the station in the manner originally intended, or they may modify it to meet their unique training needs.*]

Description

2-14. This station is [*describe station use and design*].

Targetry

2-15. Recommended targetry for training at this station includes [*describe*]. Targets may be moved, modified, or added, depending on available target systems or training requirements.

Controller

2-16. Controller requirements for this station include [*describe*].

Safety

2-17. Safety requirements and considerations for this station include [*describe*]. [*The unit or installation range SOP may add more stringent safety requirements.*]

Individual Tasks or Skills Trained

2-18. These may be found in the Digital Training Management System located online at https://dtms.army.mil/DTMS.

Training Support

2-19. Minimum materiel requirements to support training at this station include [describe].

Conduct of Training

2-20. Example training scenarios and sequence of events for this station include [describe]. [The UAC can accommodate an assortment of training ammunition and pyrotechnics.]

STATION 1: INDIVIDUAL AND TEAM TASK/TECHNIQUE TRAINER

2-21. This station is used to train individual and collective tasks, tactics, techniques, and procedures, specifically *Enter a Building and Clear a Room* and *Engage Targets* (Figure 2-2). It has three rooms where team leaders and squad leaders train the basics of building and room clearing.

DESCRIPTION

2-22. Station 1 has a single wood structure. Its three rooms have one or more open doorways and windows, and a man-sized hole.

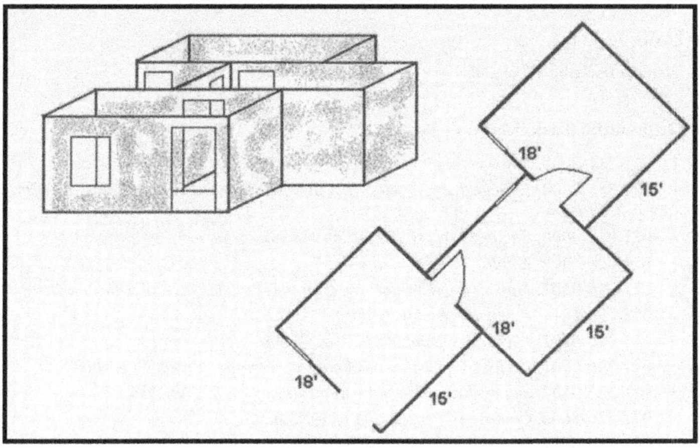

Figure 2-2. Individual and team task/technique trainer, Station 1.

TARGETRY

2-23. Targets present realistic close quarters combat situations (0 to 15 meters), which consist of a possible mixture of combatant and noncombatant targets. Target discrimination is trained by presenting up to six precision targets. Targetry sensors respond only to lethal shot placement. They also aid in training during limited visibility.

CONTROLLER REQUIREMENTS

2-24. These are based on unit requirements.

SAFETY REQUIREMENTS

2-25. Only smoke and practice grenades, or M84 stun grenades, may be used at this station.

WARNING

To avoid injury, place the weapon on SAFE when climbing through obstacles such as windows, fences, and walls during training.

INDIVIDUAL/COLLECTIVE TASKS TRAINED

2-26. This is noninclusive.

Collective Task

2-27. *Enter a Building/Clear a Room.*

Supporting Individual Tasks

- 071-311-2007, *Engage Targets with M16-Series Rifle/M4 Carbine Rifle.*
- 071-315-2308, *Engage Targets with M16-Series Rifle/M4 Carbine Using a Night Vision Sight AN/PVS-4.*
- 071-010-0006, *Engage Targets with M249 Machine Gun.*
- 071-325-4407, *Employ Hand Grenades.*
- 071-326-0503, *Move Over, Through, or Around Obstacles (except Minefields).*
- 052-193-1013, *Neutralize Booby Traps.*
- 071-326-0501, *Move as a Member of a Fire Team.*
- 071-326-0541, *Perform Movement Techniques during an Urban Operation.*
- 071-326-0557, *Select Hasty Firing Positions during an Urban Operation.*
- 071-326-5605, *Control Movement of a Fire Team.*
- 071-326-5611, *Conduct the Maneuver of a Squad.*
- 071-710-0008, *Operate Night Vision Goggles AN/PVS-7B.*

Related Tactics, Techniques, and Procedures

- FM 3-06.11, *Combined Arms Operations in Urban Terrain.*
- FM 3-22.9, *Rifle Marksmanship.*

TRAINING SUPPORT REQUIREMENTS

2-28. Use Table 2-1 for the minimum ammunition required to train at this station.

Table 2-1. Support requirements for Station 1.

Ammunition	*Rounds per Weapon*
5.56-mm Blank	10 rounds
5.56 mm Blank (Linked)	15 rounds
Practice Grenade Bodies	2 per team
Practice Grenade Fuses	2 per team

CONDUCT OF TRAINING

2-29. Table 2-2 lists the training sequence and times for Station 1.

Table 2-2. Sequence of events and times for Station 1.

Event	*Action*	*Estimated Time*
1	*Employ Hand Grenades*	30 minutes crawl, walk, run
	Perform Movement Techniques • *Engage Target with M16* • *Target Discrimination*	20 minutes crawl 20 minutes walk 10 minutes run
2	*Enter a Building*	20 minutes crawl 20 minutes walk 10 minutes run
3	*Enter and Clear a Room*	20 minutes crawl 20 minutes walk 10 minutes run
4	*Conduct an AAR*	40 minutes

Example Scenarios, Assault Team, Varied Entry Positions

2-30. Figures 2-3 and 2-4 show example scenarios. Rather than using the examples, units should tailor their training events based on their unit's METL and training needs.

Varied Entry Positions

2-31. The team leader first trains to execute *Varied Entry Positions* without targets, which is part of the *crawl* phase of training. The trainer has the assault teams position at various places to demonstrate how the start position can affect the final points of domination. (Figure 2-3).

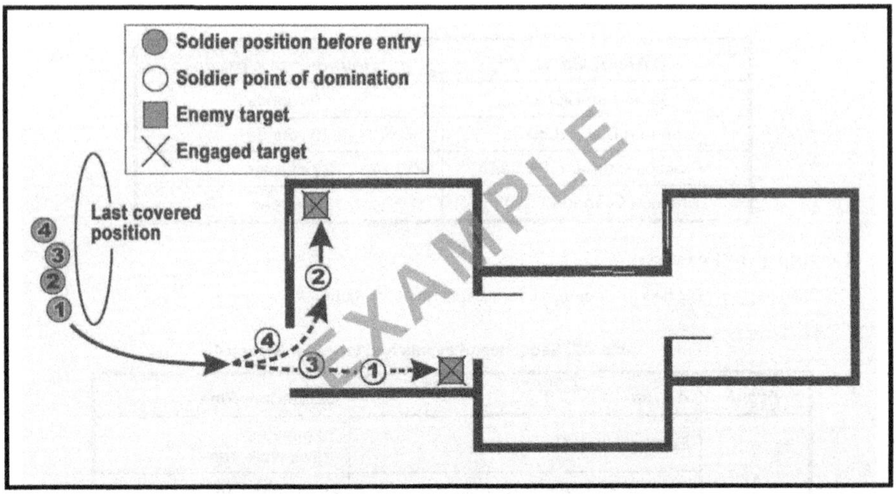

Figure 2-3. Example Station 1 scenario, *Varied Entry Positions*.

Multiple Room-Clearing Techniques

2-32. The team leader trains his Soldiers to execute *Varied Entry Positions* using targets in a multiple room scenario. This is part of the *walk* phase. The trainer has the assault teams enter the room and engage targets. He has them use using precision marksmanship techniques with the Tactical Engagement System (TES) or SESAMS (Figure 2-4). The assault team enters the first room and engages the targets. As the number two Soldier clears his area of responsibility, he sees and engages the target in the second room, but remains in place until the assault teams finishes clearing the first room. Once the room is clear, the assault team moves to the second room and continues clearing, or lets the next team move through.

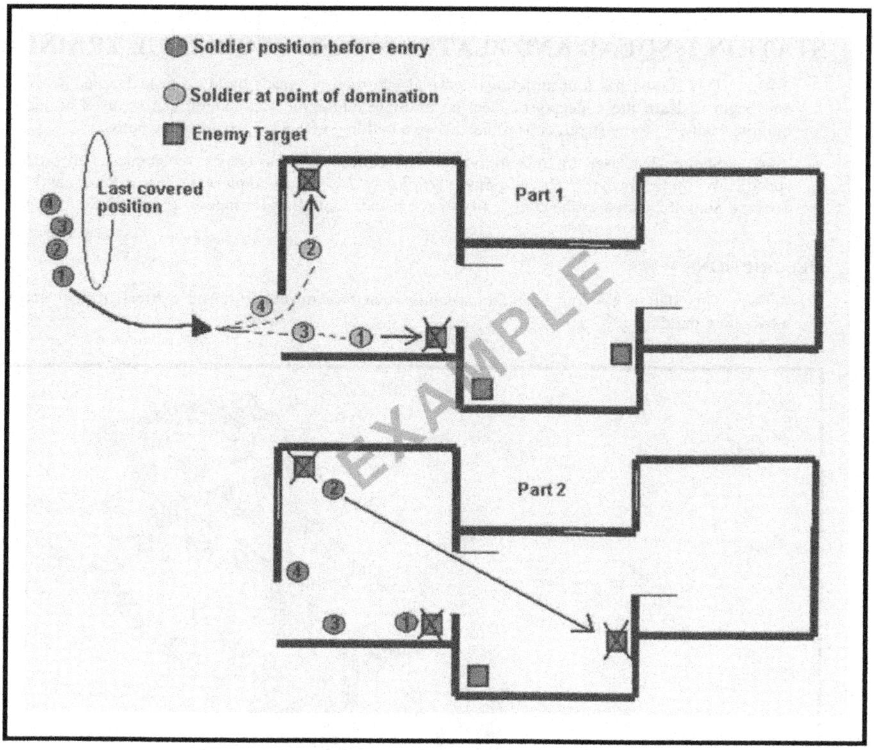

Figure 2-4. Example Station 1 scenario,
Multiple Room-Clearing Techniques.

STATION 2: SQUAD AND PLATOON TASK/TECHNIQUE TRAINER

2-33. This station has four structures and multiple rooms. Squads build on tasks learned at Station 1, and begin to learn the concepts of clearing multiple buildings. This station can be used as individual buildings with a narrow street, or as rooms inside a building with a long, connecting hallway.

2-34. Station 2 is used to train individual and collective tasks, tactics, techniques, and procedures, specifically, *Move Tactically Down a Street (Hallway)*, *Enter a Building and Clear a Room*, and *Engage Targets*. Station 2 increases the complexity of command, control, and maneuver (Figure 2-5).

DESCRIPTION

2-35. This station has four separate structures that face inward to create a street. It also introduces a two-story building.

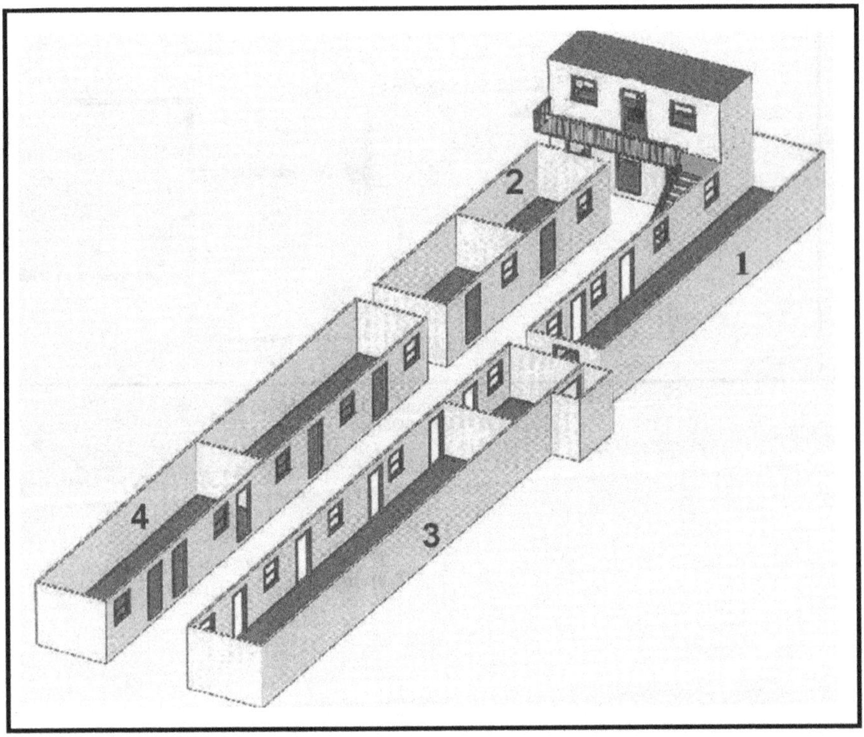

Figure 2-5. Squad and platoon task and technique trainer, Station 2.

Building 1 is L-shaped.

- The long leg of the building to the right looks like a single-story building.
- The short leg of the building to the front has two floors. A landing along the length of the second floor has outside stairs. An inside stairway also joins the floors.
- The two legs of the "L" are adjacent, but without interior access to each other.

Building 2 has a single floor and two rooms, each with a door and window. An open area much like a street separates this structure from the first.

Building 3 is a row of wooden buildings along a street. It has many windows and doorways. At the far end is an adjoining room.

Building 4 is on the left. It mirrors Building 3, plus it has a wall on the side at the near end.

2-36. Units can place additional walls in any structure by using target cloth or other material. They can do this to add to the number of rooms or to increase the complexity of the training (Appendix A).

TARGETRY

2-37. Station 2 targets present realistic close quarters combat situations (0 to 15 meters), with the option to customize the mixture of combatant and noncombatant targets. Target discrimination is trained on up to ten precision targets. Targetry sensors respond only to lethal shot placement, and facilitate training during limited visibility.

CONTROLLER REQUIREMENTS

2-38. These are based on unit requirements.

SAFETY REQUIREMENTS

2-39. At this station, Soldiers may only use smoke grenades, practice grenades, and M84 stun grenades.

WARNING

To avoid injury, place the weapon on SAFE when climbing through obstacles such as windows, fences, and walls during training.

INDIVIDUAL AND COLLECTIVE TASKS TRAINED

2-40. This list is noninclusive.

Collective Task--*Enter and Clear a Building*

Related Collective Tasks

- *Conduct Consolidation and Reorganization (Company/Platoon).*
- *Treat Casualties.*
- *Evacuate Casualties.*
- *Defend MOUT Building.*

- *Conduct Tactical Movement.*
- *Enter and Clear a Room.*

Supporting Individual Tasks

- 052-193-1013, *Neutralize Booby Traps.*
- 071-010-0006, *Engage Targets with M249 Machine Gun.*
- 071-311-2007, *Engage Targets with M16-Series Rifle/M4 Carbine Rifle.*
- 071-315-2308, *Engage Targets with M16-Series Rifle/M4 Carbine Using a Night Vision Sight AN/PVS-4.*
- 071-325-4407, *Employ Hand Grenades.*
- 071-326-0501, *Move as a Member of a Fire Team.*
- 071-326-0503, *Move Over, Through, or Around Obstacles (except Minefields).*
- 071-326-0541, *Perform Movement Techniques during an Urban Operation.*
- 071-326-0557, *Select Hasty Firing Positions during an Urban Operation.*
- 071-326-0608, *Use Visual Signaling Techniques.*
- 071-326-5605, *Control Movement of a Fire Team.*
- 071-326-5611, *Conduct the Maneuver of a Squad.*
- 071-710-0008, *Operate Night Vision Goggles AN/PVS-7B.*
- 113-571-1022, *Perform Voice Communications.*
- 191-377-5250, *Handle Enemy Personnel and Equipment.*

Related Tactics, Techniques, and Procedures

- FM 3-06.11, *Combined Arms Operations in Urban Terrain.*
  -- Chapter 3, *Offensive Operation.*
  -- Chapter 5, *Fundamental Combat Skills.*
  -- Appendix M, *Field-Expedient Breaching of Common Urban Barriers.*
- FM 3-22.9, *Rifle Marksmanship.*

TRAINING SUPPORT REQUIREMENTS

2-41. Table 2-3 shows the minimum ammunition required to train at this station.

Table 2-3. Support requirements for Station 2.

Ammunition	*Rounds per Weapon*
5.56-mm blank	16
5.56-mm blank (linked)	24
Practice grenade fuses	8 per squad
Practice grenade bodies	8

CONDUCT OF TRAINING

2-42. Table 2-4 shows the sequence of events and times for Station 2. Figure 2-6 through Figure 2-28 show example scenarios. However, rather than use these examples, the unit should customize its training events based on its METL and training needs.

Table 2-4. Sequence of events and times for Station 2.

Event	Action	Estimated Time
1	*Conduct Tactical Movement in a Built-Up Area* *Move Across an Open Area* *Move Parallel to Buildings*	10 min Crawl 20 min Walk 10 min Run
2	*Enter and Clear a Room (Squad)* *Clear a Stairwell* *Clear a Hallway* *Conduct a Breach*	10 min Crawl 20 min Walk 10 min Run
3	*Enter and Clear a Building (Platoon)* *Clear Multiple Rooms* *Consolidate and Reorganize*	10 min Crawl 20 min Walk 10 min Run
4	*Conduct AAR (Squad)*	10 min
NOTE: Training should include techniques outlined FM 3-06.11.		

2-43. The squad is clearing Station 2 as if it were a single building, and the street a hallway. Fire Team A moves down the left side, throws the appropriate grenade IAW ROE, enters, engages the target, and clears the room on that side. The last Soldier remains outside and secures the rear Fire Team B prepares to enter the next room (Figure 2-6).

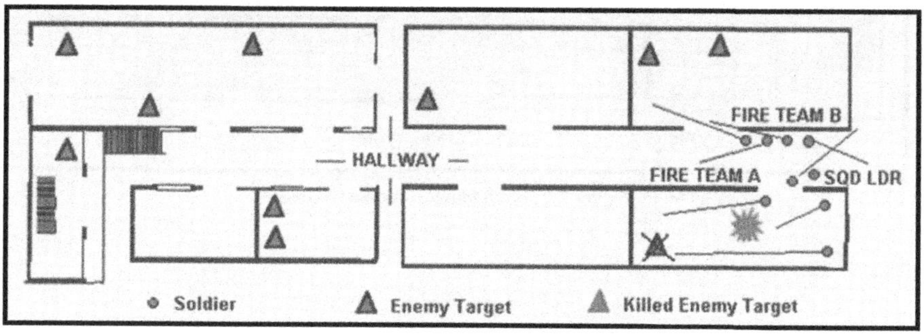

Figure 2-6. Conducting tactical movement.

2-44. Fire Team B throws the appropriate grenade, enters, engages the targets, and clears the room on the right while the last Soldier secures the rear Fire Team A prepares to continue movement to the next room (Figure 2-7).

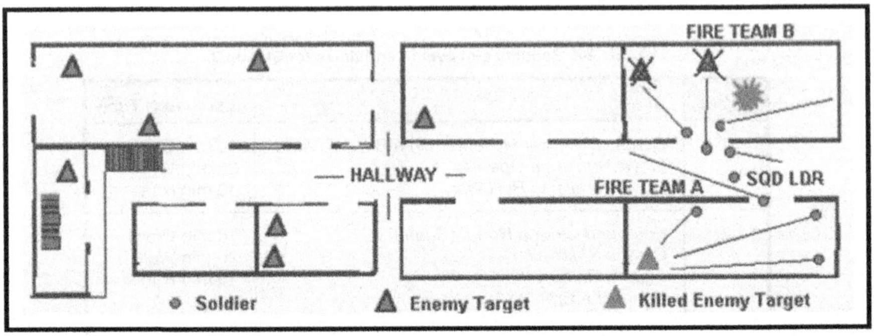

Figure 2-7. Team B--Clearing a building.

2-45. Fire Team A moves across the hall, throws the appropriate grenade, enters, engages the target, and clears the next room Three team members enter the room, the fourth remains in the doorway and secures the hallway (Figure 2-8).

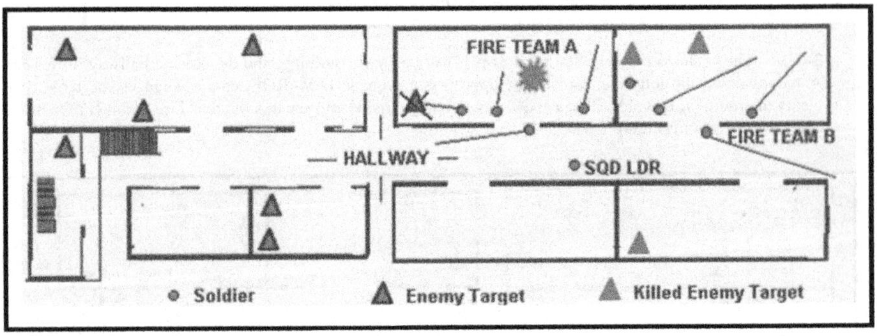

Figure 2-8. Team moving across the hall.

2-46. Fire Team B moves in a cross-cover formation down the hallway to the next room, throws the appropriate grenade, and clears the room. As Fire Team B passes the security man of Team A, he turns and becomes rear security. The last Soldier in Fire Team B remains by the door and secures the hallway (Figure 2-9).

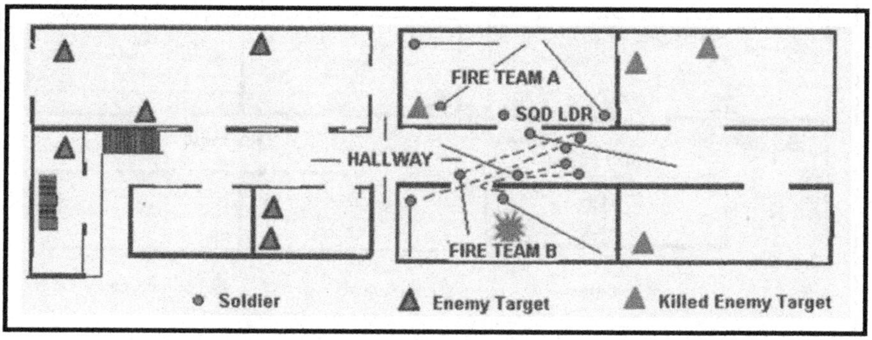

Figure 2-9. Team B--Moving in cross cover formation.

2-47. Fire Team A clears the hallway intersection, while Team B uses the Rolling-T technique to pass through Team A en route to the next room (Figure 2-10).

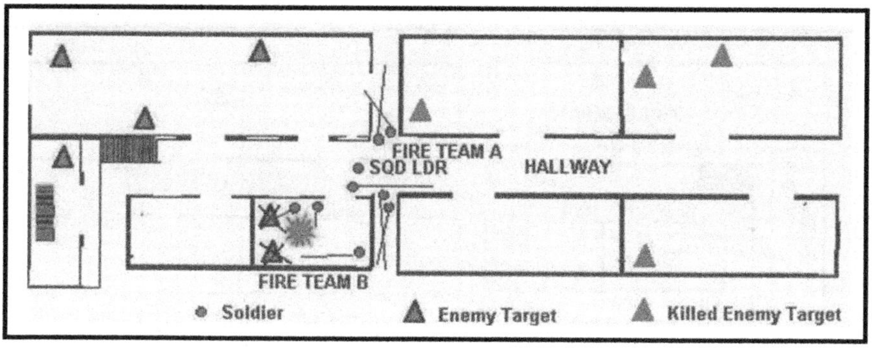

Figure 2-10. Team A--Clearing a hallway intersection.

2-48. Fire Team B throws the appropriate grenade, enters, engages the targets, and clears the room. Only three Soldiers enter, while the fourth Soldier provides rear security. Fire Team A prepares to move (Figure 2-11).

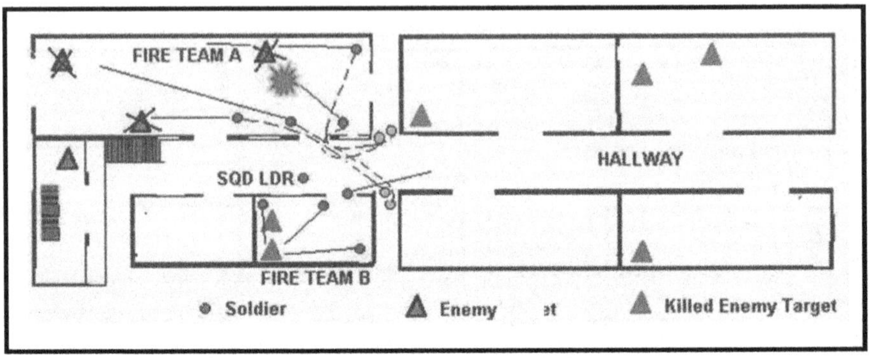

Figure 2-11. Team B--Clearing a room.

2-49. Fire Team A moves from the intersection and throws the appropriate grenade. The team enters, engages the targets, and then clears the next room. All team members enter, because of the size of the room. Fire Team B prepares to move (Figure 2-12).

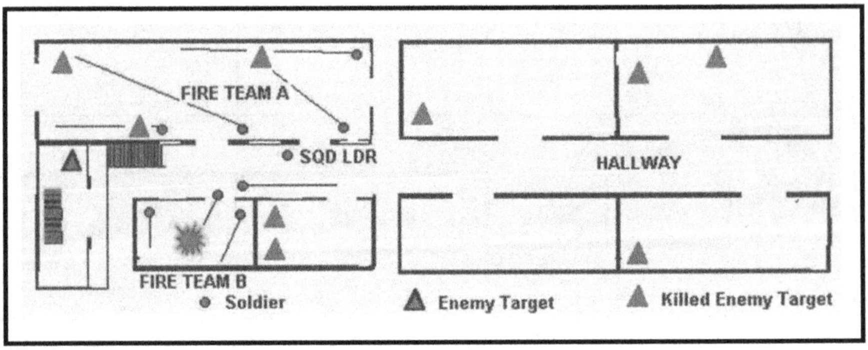

Figure 2-12. Team A--Clearing room by room.

2-50. Fire Team B employs the appropriate grenade, enters, and engages the target. Three Soldiers clear the next room. The fourth remains in the doorway and secures the rear. Fire Team A prepares to move up the stairs (Figure 2-13).

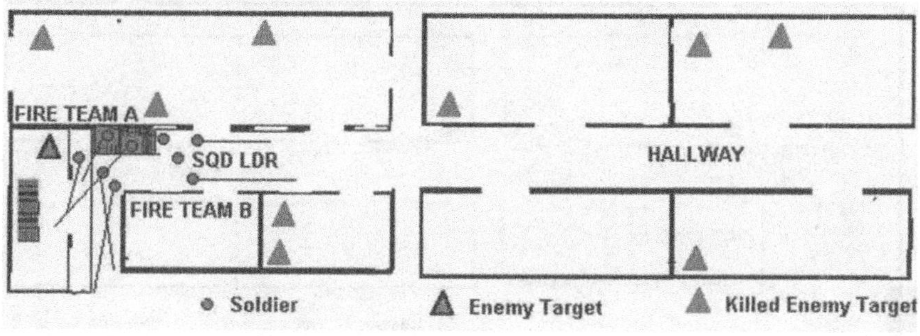

Figure 2-13. Team B--Clearing the next room.

2-51. Fire Team A moves up the stairs and onto the landing. The team prepares to enter the room off the landing. Fire Team B splits, and two members move around the corner and secure the area across from the stairs on the lower level. The other two members of Team B secure the hallway the squad came from (Figure 2-14).

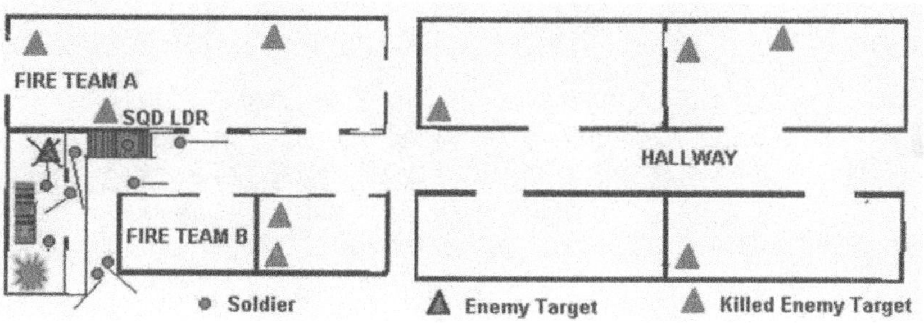

Figure 2-14. Securing the stairs and hallway.

2-52. Fire Team A members employ the appropriate grenade, enters the room, engages the target, and clears it. They find another set of stairs, and prepare to move up them. Fire Team B continues to secure the rear. The other team members move to the end of the short hallway and secure both directions (Figure 2-15).

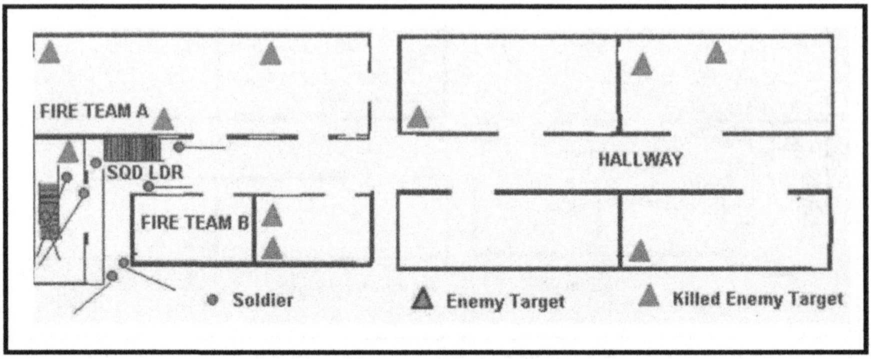

Figure 2-15. Preparing to clear more stairs.

2-53. Fire Team A clears the stairs and the landing. Fire team B continues to secure the ground level, ready to assist Team A, if necessary (Figure 2-16).

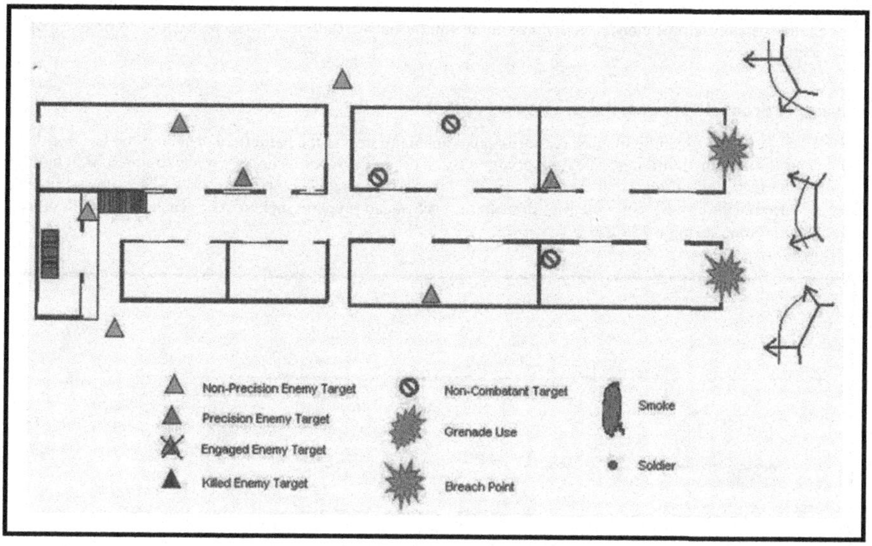

Figure 2-16. Team A--Clearing stairs and landing.

2-54. The previous example used no noncombatant targets, but used blanks/SESAMS and TES. This technique is the *crawl* phase for this station. Once the squad is ready to move on to more challenging exercises, the platoon leader/sergeant can add targets into the scenario. As proficiency increases, mixed combatant and noncombatant targets can be added. At any point, the platoon leader/sergeant can, until the platoon achieves proficiency, include multiple squads on the station. This increases the complexity of the scenarios.

Example Scenario--Preparing to Gain a Foothold

2-55. In this example platoon scenario, noncombatants are on the battlefield. The platoon has decided to clear through the buildings. They are going to try to do this without using the streets to move from building to building. Two squads will assault, and the other will remain in support/reserve. The support element isolates the buildings, while the assault team for each squad prepares and breaches each building in order to gain a foothold (Figure 2-17).

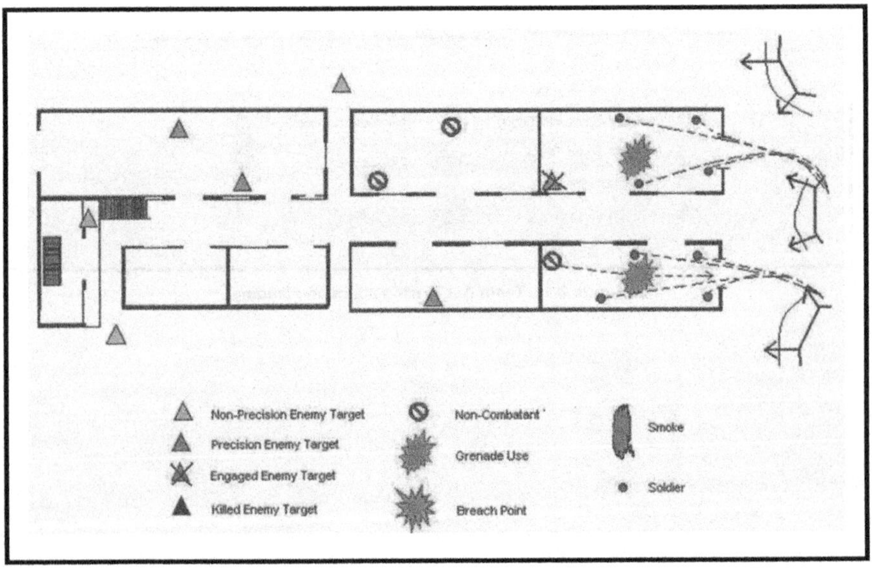

Figure 2-17. Preparing to gain a foothold.

2-56. The assault teams throw a grenade into the breach and their members follow, engaging enemy targets. They secure and move noncombatants to the collection point (CP) (Figure 2-18).

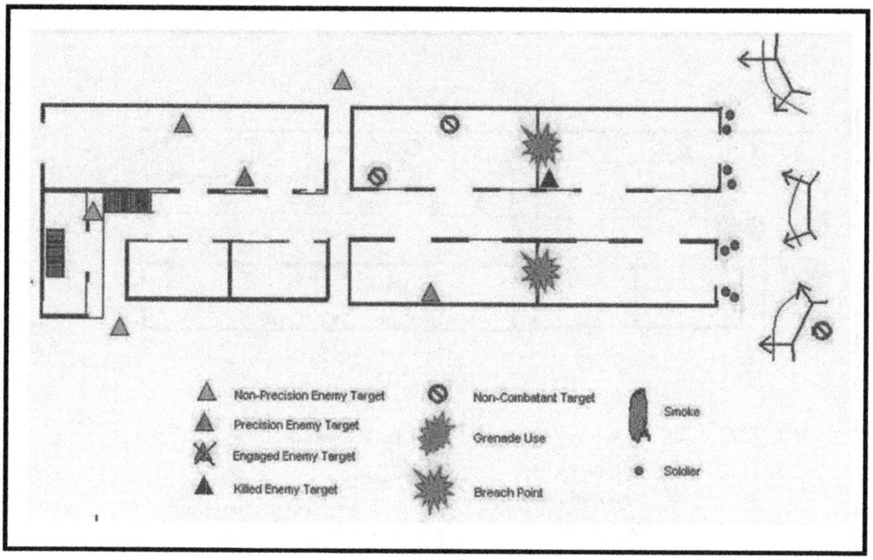

Figure 2-18. Securing and moving noncombatants to the command post.

2-57. The teams prepare and detonate wall breaches to continue movement through the building. Once the charges are ready for detonation, the assault teams move to a covered position and execute the breach (Figure 2-19).

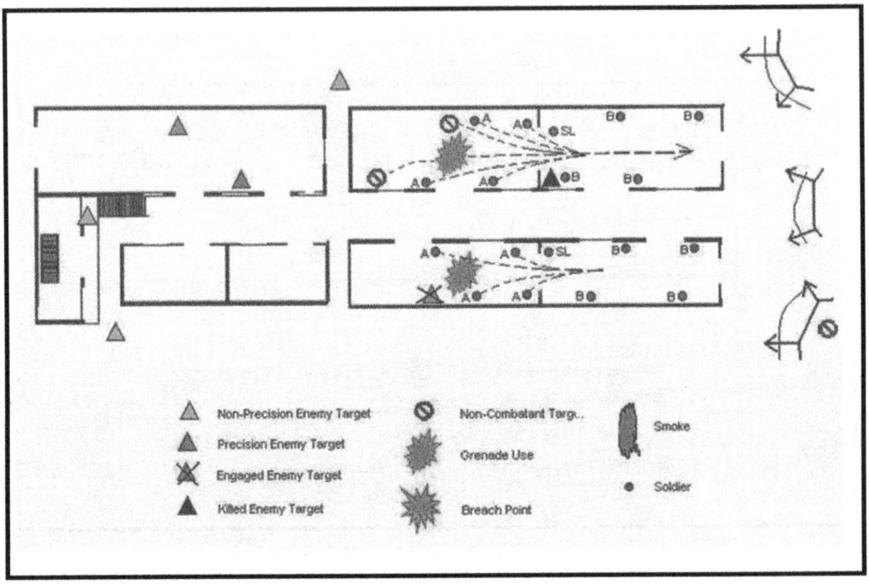

Figure 2-19. Executing breach.

2-58. The assault team employs grenades into the next room, enters the room, eliminates the enemy, and then secures and moves the noncombatants to a designated safe area. At the same time, the next assault team moves into the foothold (Figure 2-20).

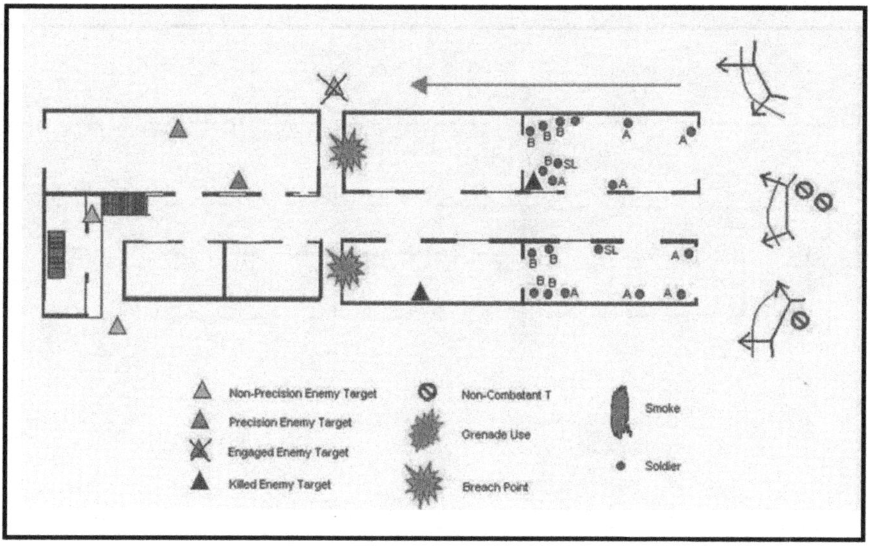

Figure 2-20. Moving to the foothold.

2-59. With the second room secured, another assault team moves forward and prepares to breach the wall Once the charges are placed, the teams move back to a covered position before detonating the charge. With the breach a success, the first nonprecision target is exposed and engaged by the support element (Figure 2-21).

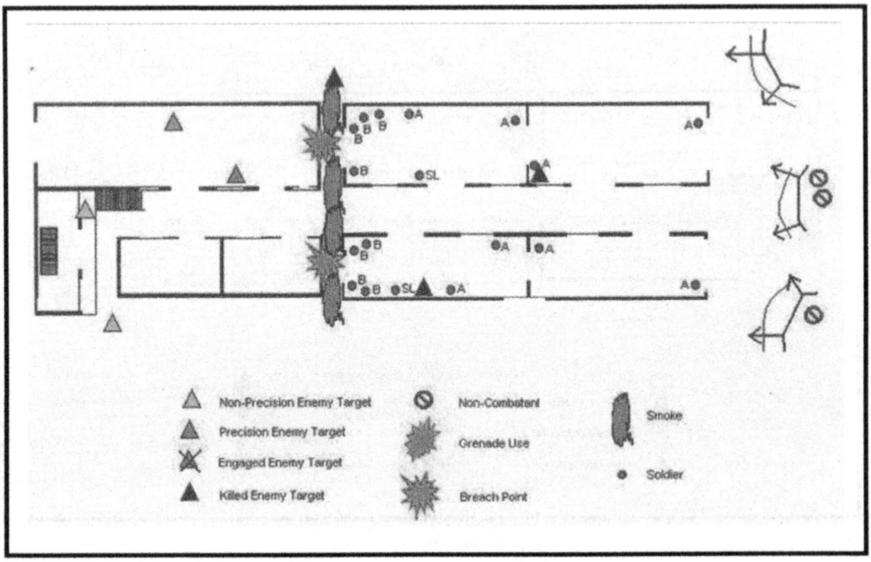

Figure 2-21. Breaching the wall.

2-60. Upon inspecting the breach, the assault teams find that they have an alley with another wall to breach. They employ smoke to obscure the teams placing the breaching charge. The first assault teams take up securing positions throughout the already cleared areas. Follow-on assault teams move to a covered position and execute the breach (Figure 2-22).

Note: Use smoke only in the alley, not inside the buildings.

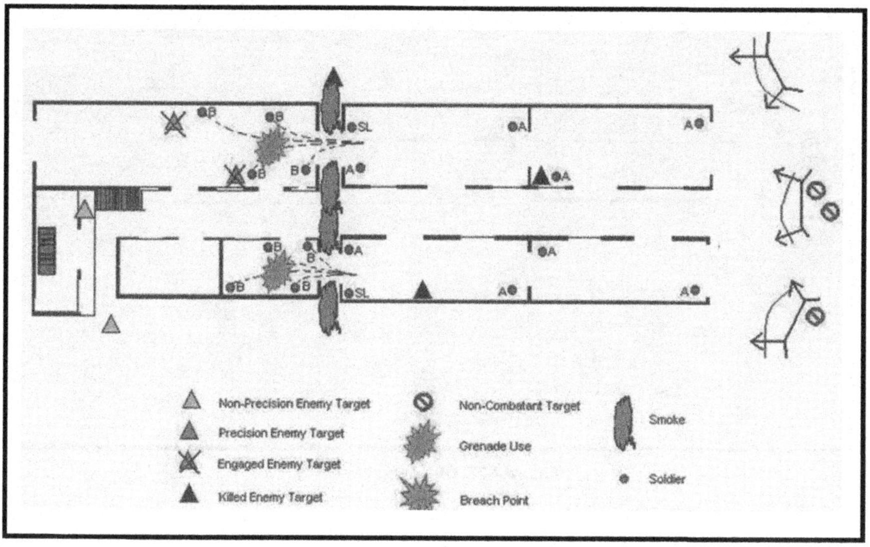

Figure 2-22. Securing a wall in an alley.

2-61. Once the assault teams execute the breach, they must determine if they need to use more smoke to cover their movement. Either way, once they have enough smoke to cover their movement, they throw in a grenade, enter, engage enemy targets, and clear the rooms (Figure 2-23).

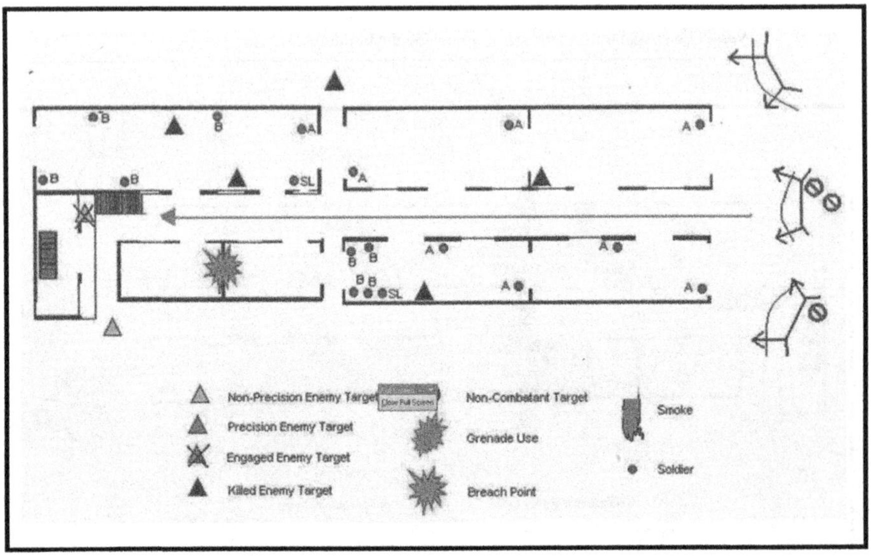

Figure 2-23. Obtaining obscurity.

2-62. While one team finds itself at the end of the building in their sector, the other team is faced with another wall and more clearing. One team prepares to breach the wall, and the other team takes up a hasty position and prepares for a counterattack. When the teams execute the breach, the next nonprecision target presents, and the support element engages it (Figure 2-24).

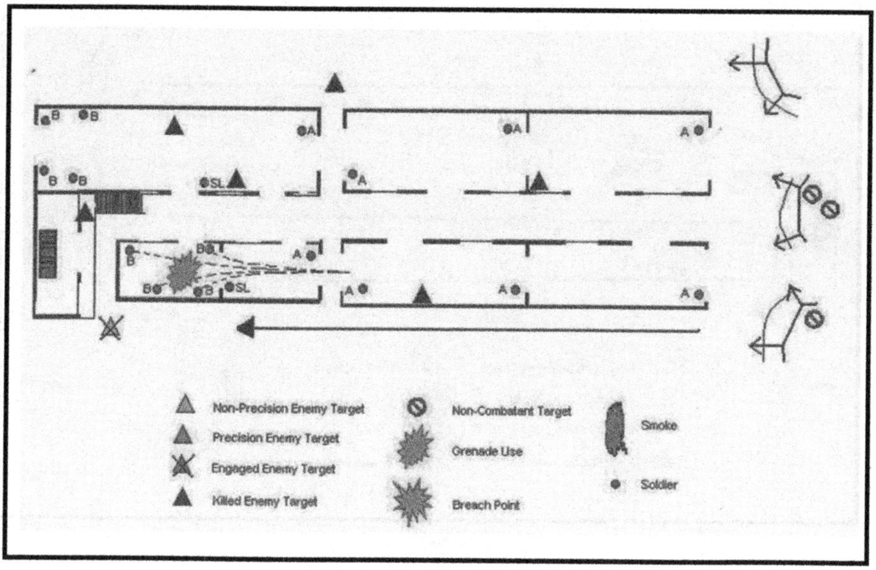

Figure 2-24. Clearing the wall.

2-63. Someone throws a grenade through the breach. The assault team enters and secures the room. When the grenade explodes inside the room, the final nonprecision target presents, and the support element engages it (Figure 2-25).

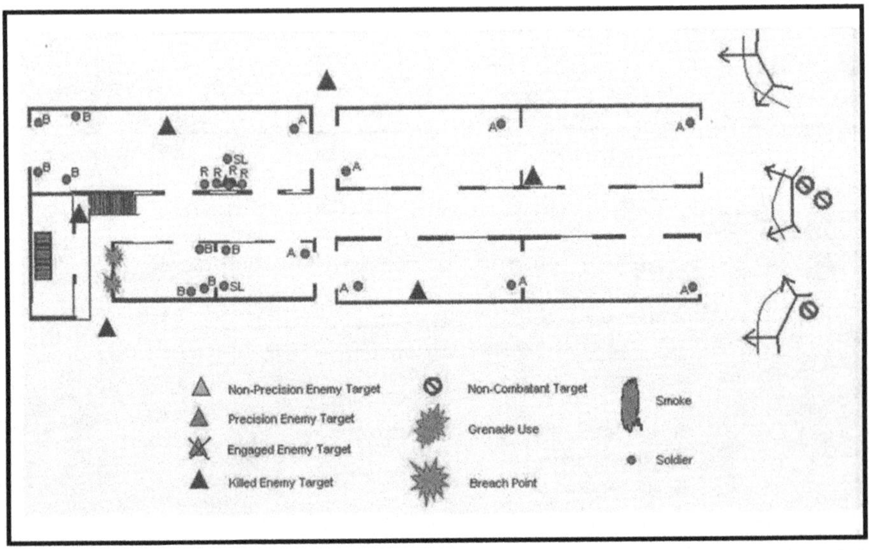

Figure 2-25. Securing a room.

2-64. With the first squad reaching its limit of advance, and with the second squad reaching the end of the building, the platoon leader calls another assault team forward from the reserve. The reserve assault team moves forward through the buildings, and positions itself to clear up the outside stairs of the two-story structure. The second squad places and detonates loophole charges on the far wall of the building. This supports the clearing of the last building (Figure 2-26).

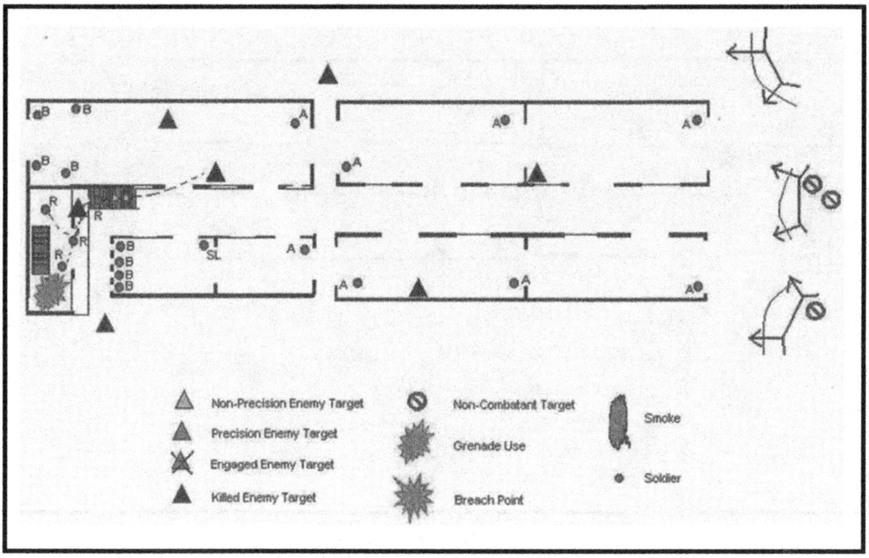

Figure 2-26. Reserve assault team moving forward.

2-65. Team B of the second squad moves into supporting fire positions. The new assault teams move up the outside stairs to the second floor of the final structure. They clear the balcony and the first room on the second floor. The support-by-fire position shift fires to the lower floor of the same structure (Figure 2-27).

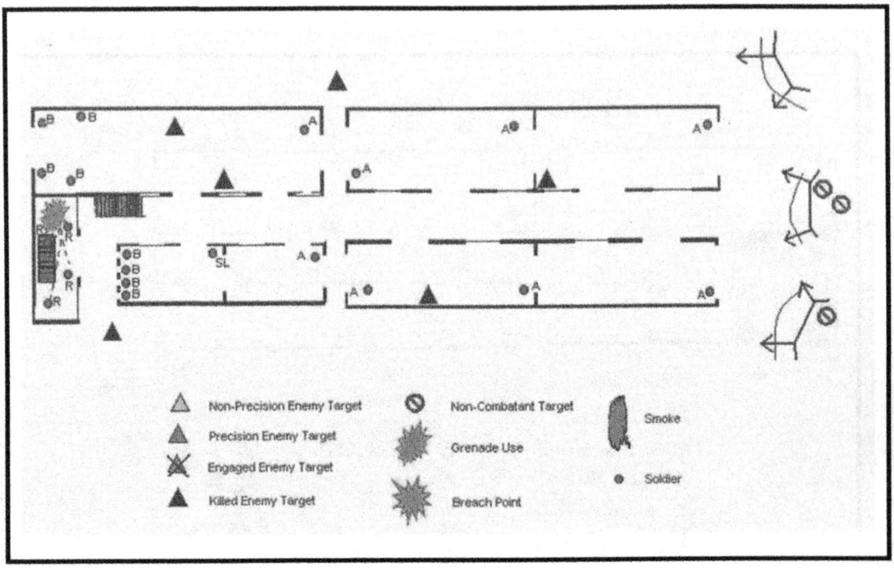

Figure 2-27. Team B, second squad--moving into the supporting fire position.

2-66. Once the floor is secure, the assault team moves down the stairs and clears the remaining floor (Figure 2-28). Meanwhile, the platoon repositions, and then consolidates and reorganizes.

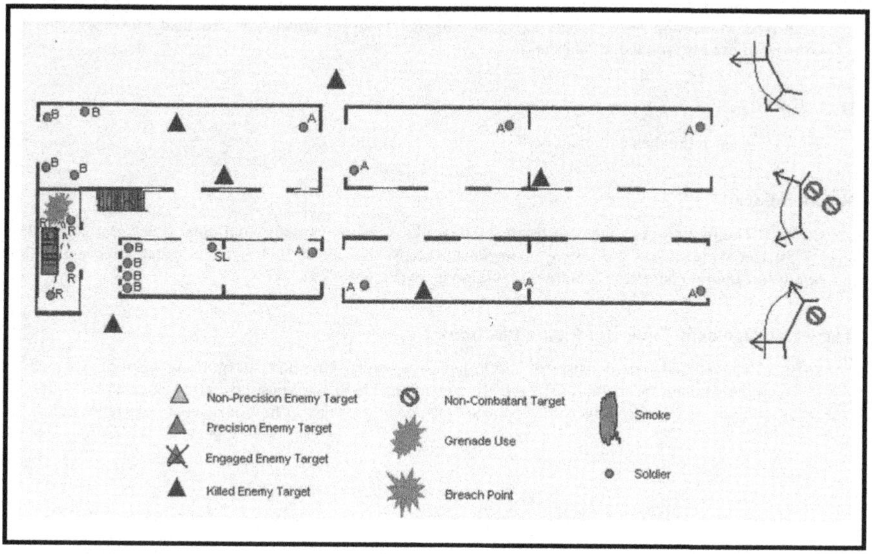

Figure 2-28. Consolidation and reorganization.

STATION 3: GRENADIER GUNNERY TRAINER

2-67. This is a live-fire station where M203 dual-purpose weapon gunners master the engagement of point-type M203 and M16 targets at 50 to 150 meters in an urban area. Tactical movement and fire commands are also trained at this station.

DESCRIPTION

2-68. This station has two main parts:

Wooden Facade

2-69. The first part is a wooden facade wall that looks like a two-story building. It is about 20 feet high by 30 feet wide. It has a doorway with outside steps, two lower floor windows and three upper floor windows, and two basement windows at ground level (Figure 2-29).

Target Engagement Lane and Firing Positions

2-70. The second part consists of the target engagement lane and seven firing positions, which are between the wooden façade and the first firing position. These positions have cover such as sandbags, log walls, and rubble. Firing points can be adjusted to support a variety of training scenarios (Figure 2-30).

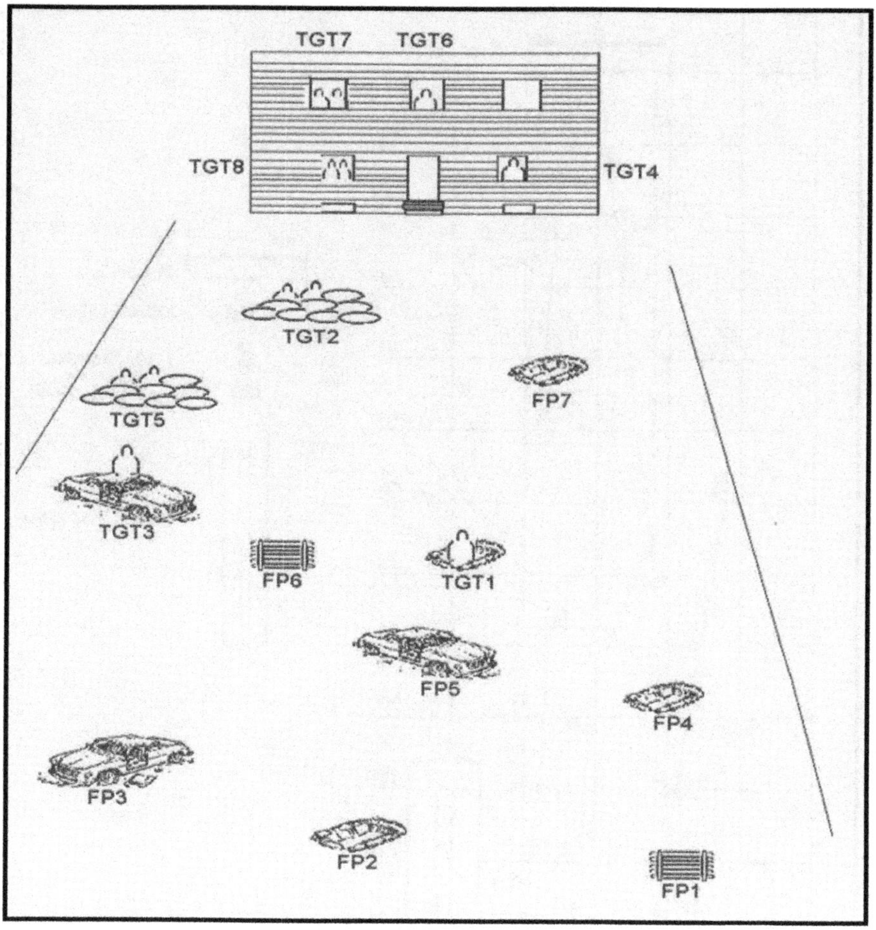

Figure 2-29. Part 1, Grenadier Gunnery Trainer, Station 3, wooden facade.

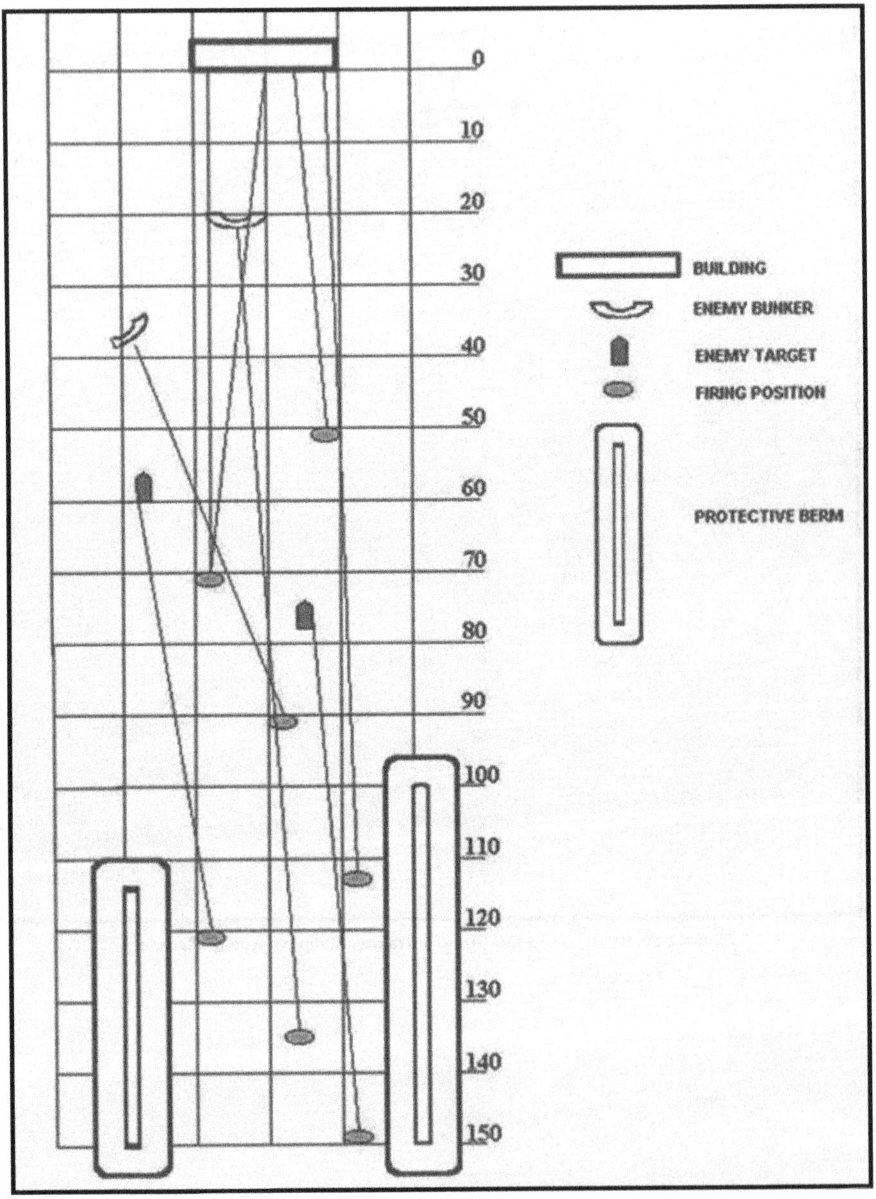

Figure 2-30. Part 2, Grenadier Gunnery Trainer, Station 3, target engagement lane and firing positions.

WARNING

Never engage targets with the 40-mm at ranges less than 31 meters, which is the minimum safe range. Trainers ensure the station accommodates this requirement as well as the 14-meter arming range.

TARGETRY

2-71. This station has twelve target mechanisms. These mechanisms are arranged so as to influence the gunner's decision as to which munition to use.

CONTROLLER REQUIREMENTS

2-72. As required by unit.

SAFETY REQUIREMENTS

2-73. Check the unit SOP and local safety regulations regarding live-fire ranges. Ensure all weapon systems are on SAFE before moving to the next firing position. Ensure that no 40-mm target is engaged at less than minimum arming range.

INDIVIDUAL AND COLLECTIVE TASKS TRAINED

2-74. These lists are noninclusive:

Supporting Individual Tasks

- 071-311-2007, *Engage Targets with M16-Series Rifle/M4 Carbine Rifle.*
- 071-311-2130, *Engage Targets with M203 Grenade Launcher.*
- 071-326-0503, *Move Over, Through, or Around Obstacles (except Minefields).*
- 071-326-0541, *Perform Movement Techniques during an Urban Operation.*
- 071-326-0557, *Select Hasty Firing Positions during an Urban Operation.*
- 071-326-0608, *Use Visual Signaling Techniques.*

Related Tactics, Techniques and Procedures

- FM 3-06.11, *Combined Arms Operations in Urban Terrain.*
  -- Chapter 3, *Offensive Operations.*
  -- Chapter 5, Fundamental Combat Skills.
- Appendix M, *Field Expedient Breaching of Common Urban Barriers.*

Training Support Requirements

2-75. Table 2-5 shows the minimum materials required to train at this station.

Table 2-5. Support requirements for Station 3.

Ammunition	*Rounds for each Weapon*	*Targets Required*
5.56-mm Ball	8	Four nonprecision human target mechanisms.
40-mm TP	8	Eight nonprecision human target mechanisms.

Conduct of Training

2-76. Whereas Table 2-5 showed the minimum amount of ammunition required for this station, Table 2-6 shows an example sequence of events and times, and Table 2-7 shows an example target scenario.

Table 2-6. Sequence of events and times for Station 3.

Event	*Action*	*Estimated Time*
1	*Perform Movement Techniques* *Select Hasty Firing Position* *Engage Targets With 5.56-mm (dry run)* *Engage targets with 40mm*	20 min
2	*Perform Movement Techniques* *Select Hasty Firing Position* *Engage Targets with 5.56mm (live-fire)* *Engage targets with 40mm*	10 min
NOTE: Training should include techniques outlined in FM 3-06.11		

Table 2-7. Example target scenario (presentation order) for Station 3.

Firing Position	*Weapon*	*Presentation Time*	*Range to Target*	*Target Presented*
1	5.56mm	6 sec	67 meters	Single Human Target (TGT 1)
2	40mm	2 min	105 meters	Multiple Human Targets (TGT 2)
3	5.56mm	6 sec	63 meters	Single Human Target (TGT 3)
4	5.56mm	6 sec	105 meters	Single Human Target (TGT 4)
5	40mm	2 min	57 meters	Multiple Human Targets (TGT 5)
6	40mm	2 min	70 meters	Multiple Human Targets (TGT 6)
6	5.56mm	6 sec	70 meters	Single Human Target (TGT 8)
7	40mm	2 min	50 meters	Multiple Human Targets (TGT 7)

STATION 4: URBAN OFFENSE/DEFENSE BUILDING, TASK/TECHNIQUE TRAINER

2-77. This station trains to attack or defend a building. Collective and individual TTP are trained. The platoon practices the task steps and performance measures for offensive and defensive tasks. This station increases the complexity of command, control, and maneuver (Figure 2-31).

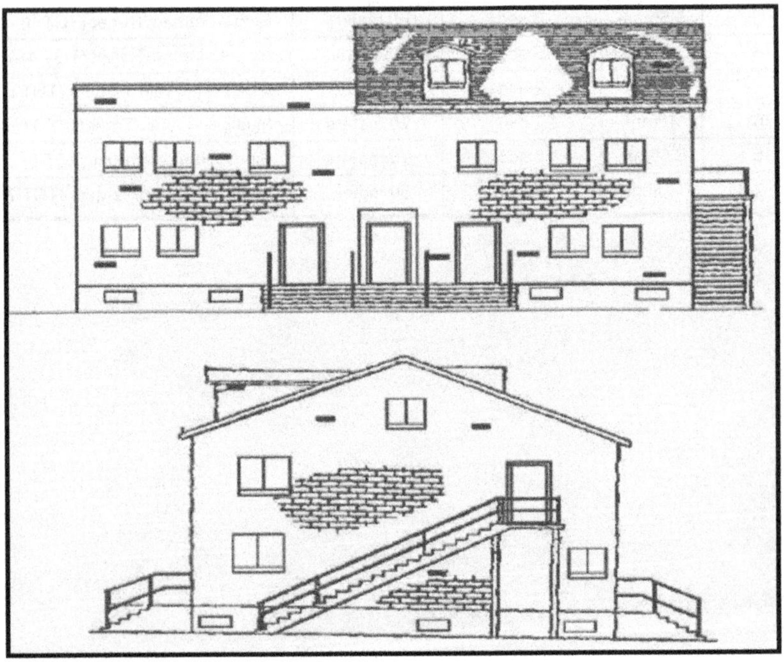

Figure 2-31. Urban offense/defense building, Station 4.

DESCRIPTION

2-78. This station is a two-story building with a basement. Each floor has several rooms, windows, doorways, loopholes, and mouseholes. The gabled roof has windows and a hatch to the attic below and half of the rooftop is flat. The building has exterior and interior stairways.

TARGETRY

2-79. This station has ten target mechanisms.

CONTROLLER REQUIREMENTS

2-80. These are based on unit requirements.

SAFETY REQUIREMENTS

2-81. Only smoke, practice, and M84 stun grenades may be used at this station. It is not a live-fire station. Training is conducted using TES with blank 5.56-mm, or TES or SESAMS. Units should consult with their supporting combat engineers for the proper demolition effects simulators (DES) to use against the blow panels throughout the station.

DANGER

TO AVOID ASPHYXIATION, NEVER DETONATE SMOKE GRENADES INSIDE BUILDINGS. THEY ARE *ALL* DANGEROUS IN AN ENCLOSED SPACE.

WARNINGS

1. **To avoid injury, place your weapon on SAFE when climbing through obstacles such as windows, fences, and walls.**
2. **To avoid inflicting carbon burns or more serious injury, never fire blank ammunition at other Soldiers within 1 meter.**

DEMOLITION EFFECTS SIMULATORS

2-82. Demolition effects simulators (DES) replicate breaching to gain entry to buildings and rooms. The walls of the UAC, Shoot House, and CACTF facilities may be breached using wall breach charge DES. Generally, the interior and exterior walls in these facilities are made with blow panels, plasterboard, or similar material, and measure about 4 feet square.

2-83. The DES charges for use at the UAC, shoot house, and the CACTF are constructed to breach a predetermined entry point without damaging the facility.

Field-Expedient Charges

- Flex-linear shaped charge.
- Doorknob charge.
- E-silhouette charge.

Operational Charges

2-84. Unit trainers must consult with installation safety and combat engineers to determine the appropriate reduced charge. See also FM 3-06.11.

INDIVIDUAL OR COLLECTIVE TASKS TRAINED

Collective Task

- FM 3-21.8 (7-8).
  - -- 07-3-9018, *Clear a Building (Platoon/Squad).*
  - -- 07-6-1216, *Defend an Urban Area (Battalion/BCT).*

Related Collective Tasks

- 07-2-5027, *Conduct Consolidation and Reorganization (Company/Platoon).*
- 08-2-0003. *Treat Casualties.*
- 08-2-0004. *Evacuate Casualties.*
- 19-3-3100, *Process Detainees at a Collecting Point (CP).*
- 07-2-1342, *Conduct Tactical Movement (Company/Platoon).*
- 07-2-1189, *Conduct a Dismounted Tactical Road March (Company/Platoon).*
- 07-2-1198, *Conduct a Mounted Tactical Road March (Company/Platoon).*
- Warrior Battle Drill 12, *Enter and Clear a Room.*

Supporting Individual Tasks

- 052-193-1013, *Neutralize Booby Traps.*
- 071-010-0006, *Engage Targets with M249 Machine Gun.*
- 071-311-2007, *Engage Targets with M16-Series Rifle/M4 Carbine Rifle.*
- 071-315-2308, *Engage Targets with M16-Series Rifle/M4 Carbine Using a Night Vision Sight AN/PVS-4.*
- 071-325-4407, *Employ Hand Grenades.*
- 071-326-0501, *Move as a Member of a Fire Team.*
- 071-326-0503, *Move Over, Through, or Around Obstacles (except Minefields).*
- 071-326-0541, *Perform Movement Techniques during an Urban Operation.*
- 071-326-0557, *Select Hasty Firing Positions during an Urban Operation.*
- 071-326-0608, *Use Visual Signaling Techniques.*
- 071-326-5605, *Control Movement of a Fire Team.*
- 071-326-5611, *Conduct the Maneuver of a Squad.*
- 071-710-0008, *Operate Night Vision Goggles AN/PVS-7B.*
- 113-571-1022, *Perform Voice Communications.*
- 191-377-5250, *Handle Enemy Personnel and Equipment.*

Related Tactics, Techniques, and Procedures

- FM 3-06.11, *Combined Arms Operations in Urban Terrain.*
  -- Chapter 3, *Offensive Operations.*
  -- Chapter 5, *Fundamental Combat Skills.*
  -- Appendix M, *Field Expedient Breaching of Common Urban Barriers*
- FM 3-22.9, *Rifle Marksmanship.*

Training Support Requirements

2-85. Table 2-8 shows the minimum ammunition required to train at this station.

Table 2-8. Ammunition requirements for Station 4.

Ammunition	*Rounds for each Weapon*
5.56-mm Blank	16
5.55-mm Blank (Linked)	24
7.62-mm Blank	100
Practice Grenade Bodies	5
Practice Grenade Fuzes	10 per platoon
M84 Stun Grenade Bodies	5
M84 Stun Grenade Fuzes	10 per platoon
Smoke Grenades	4 per platoon

Conduct of Training

2-86. Table 2-9 shows the sequence of training for Station 4. Figures 2-32 and 2-33 show example scenarios of training events that can be conducted at this station. These are only examples. The unit should develop training events based on its METL and training needs.

Table 2-9. Sequence of events and times for Station 4.

Event	*Action*	*Estimated Time*
1	*Clear a Building*	2 Hours Crawl
		1 Hour Walk
		30 Minutes Run
2	*Defend Built-up Area/Building*	1 Hours Crawl
		1 Hour Walk
		2. Hour Run
3	*Conduct AAR*	1 Hour

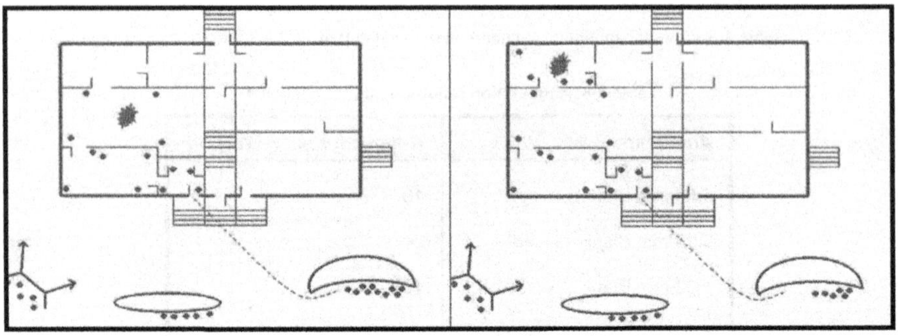

Figure 2-32. Event 1, *Clear a Building.*

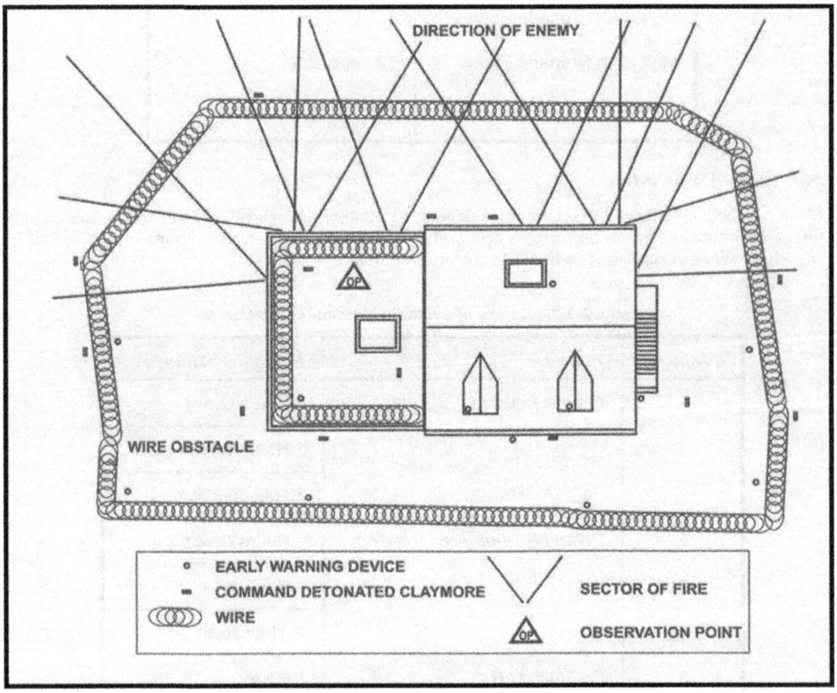

Figure 2-33. Event 2, *Defend a Building.*

STATION 5: UNDERGROUND TRAINER

2-87. This station provides training up to squad level for subterranean operations, specifically, the task *Clear and Move Tactically in a Subterranean Environment.*

DESCRIPTION

2-88. This station is a sewer system built in an irregular pattern, with four manhole covers. Two access tunnels lead from the outside tunnel to the underground loop, and the station is completely covered by dirt, only the manholes and ground level entrances are exposed (Figure 2-34).

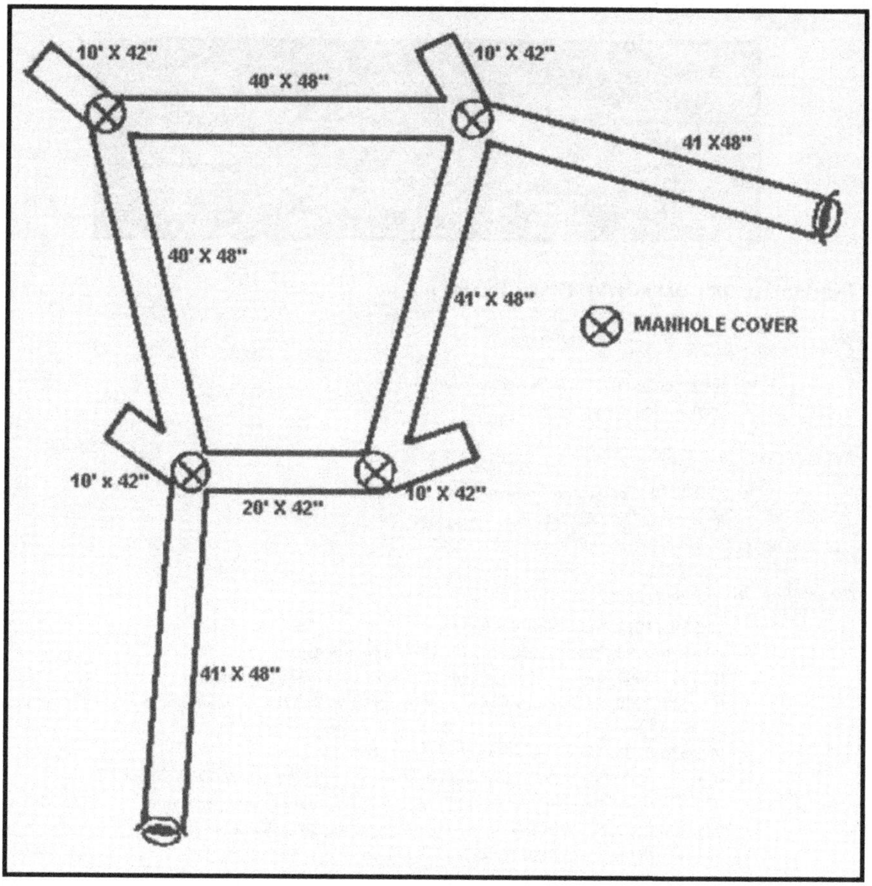

Figure 2-34. Underground Trainer, Station 5.

Targetry

2-89. Station 5 is a force-on-force training station and has no targetry.

Controller Requirements

2-90. As required by unit.

Safety Requirements

2-91. Soldiers should wear their helmets at all times when in the sewer system to avoid head injuries. Smoke grenades or M84 stun grenades *will not* be used in the underground trainer.

DANGER

TO AVOID ASPHYXIATION WHEN IN THE TRAINER, NEVER DETONATE SMOKE GRENADES IN THE SEWER. USING ANY TYPE OF EXPLOSIVE SIMULATOR IS DANGEROUS IN AN ENCLOSED SPACE. THE PROTECTIVE MASK IS NOT EFFECTIVE IN OXYGEN-DEFICIENT ATMOSPHERES.

Individual or Collective Tasks Trained

Collective Task

- *Move Tactically in a Subterranean Environment.*
- *Reconnoiter Area.*

Related Collective Tasks

- 07-2-1270, *Conduct an Infiltration (Company/Platoon).*
- 08-2-0003. *Treat Casualties.*
- 08-2-0004. *Evacuate Casualties.*

Supporting Individual Tasks

- 052-193-1013, *Neutralize Booby Traps.*
- 071-010-0006, *Engage Targets with M249 Machine Gun.*
- 071-311-2007, *Engage Targets with M16-Series Rifle/M4 Carbine Rifle.*
- 071-315-2308, *Engage Targets with M16-Series Rifle/M4 Carbine Using a Night Vision Sight AN/PVS-4.*
- 071-326-0501, *Move as a Member of a Fire Team.*
- 071-326-0503, *Move Over, Through, or Around Obstacles (except Minefields).*
- 071-326-0541, *Perform Movement Techniques during an Urban Operation.*
- 071-326-0557, *Select Hasty Firing Positions during an Urban Operation.*
- 071-326-0608, *Use Visual Signaling Techniques.*
- 071-326-5605, *Control Movement of a Fire Team.*
- 071-326-5611, *Conduct the Maneuver of a Squad.*
- 071-710-0008, *Operate Night Vision Goggles AN/PVS-7B.*

- 113-571-1022, *Perform Voice Communications.*
- 191-377-5250, *Handle Enemy Personnel and Equipment.*

Related Tactics, Techniques and Procedure

- FM 3-06.11, *Combined Arms Operations in Urban Terrain.*
  -- Chapter 3, *Offensive Operations.*
  -- Chapter 5, *Fundamental Combat Skills.*
  -- Appendix M, Field Expedient Breaching of Common Urban Barriers.
- FM 3-22.9, *Rifle Marksmanship.*

TRAINING SUPPORT REQUIREMENTS

2-92. Use Table 2-10 for the minimum ammunition required to train at this station.

Table 2-10. Ammunition requirements for Station 5.

Ammunition	Rounds for each Weapon
5.56-mm Blank	20
5.55-mm Blank (Linked)	40

CONDUCT OF TRAINING

2-93. Table 2-11 shows the sequence of training for this station. These are only examples. The unit should develop training events based on its METL and training needs.

Table 2-11. Sequence of events and times for Station 5.

Event	Action	Estimated Time
1	*Reconnoiter*	30 minutes (Crawl) 30 minutes (Walk) 30 minutes (Run)
2	*Conduct an Infiltration (Co/Plt)*	30minutes (Crawl) 30 minutes (Walk) 30 minutes (Run)
3	*Conduct AAR*	30 minutes

Chapter 3
Shoot House

This chapter describes the live-fire Shoot House and provides layouts and training scenarios. Fighting in built-up areas is fragmented; gains are often small. Therefore, units depend on the initiative, skill, and discipline of small-unit leaders and individual Soldiers. Leaders must be competent and confident in their urban operations skills and must have the courage to accomplish their missions while isolated from their parent units. Soldiers and leaders require mission-type orders that allow for decentralized execution.

PURPOSE

3-1. The Shoot House provides the leader with a facility to train and evaluate the unit during a live-fire exercise. Units are trained and evaluated on their ability to conduct tactical movement, engage targets, conduct breaches, and practice target discrimination in an urban environment.

DESCRIPTION

3-2. This facility has two parts, the live-fire facility and the AAR room:

LIVE-FIRE FACILITY

3-1. The live-fire facility is a single-level structure with a barn roof, catwalk, and eight adjoining rooms. Each room has video capability for AARs. The walls are bulletproof and prevent ricochets. Each side has an entrance. Replaceable blow panels throughout the facility can withstand reduced DES blasts. These also offer alternative entry points. A crane system in the roof allows trainers to selectively replace doorways and blow panels with solid wall sections. Before training explosive and mechanical breaching operations, trainers should place mobile exterior baffles or containment walls placed by each entry to prevent rounds and demolition effects from escaping. All of these capabilities let the unit vary its training scenarios. For safety, the Shoot House is windowless (Figure 3-1).

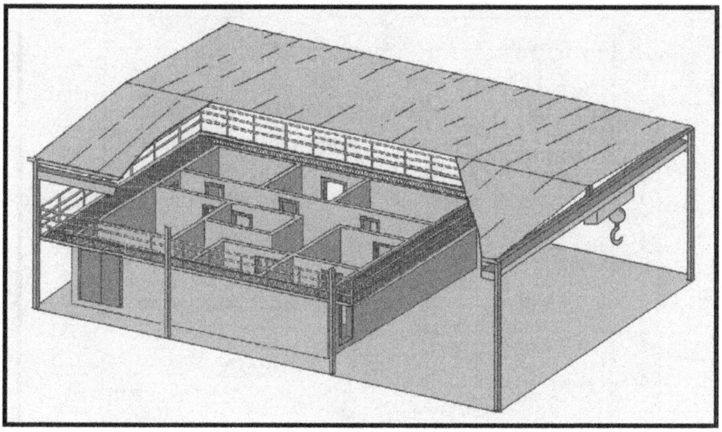

Figure 3-1. Shoot House.

AAR Building

3-2. The AAR building has a conference/AAR room that can accommodate up to a platoon size element, an instrumentation and targetry control room, and an AAR edit and projection room that can handle a formal AAR (Figure 3-2).

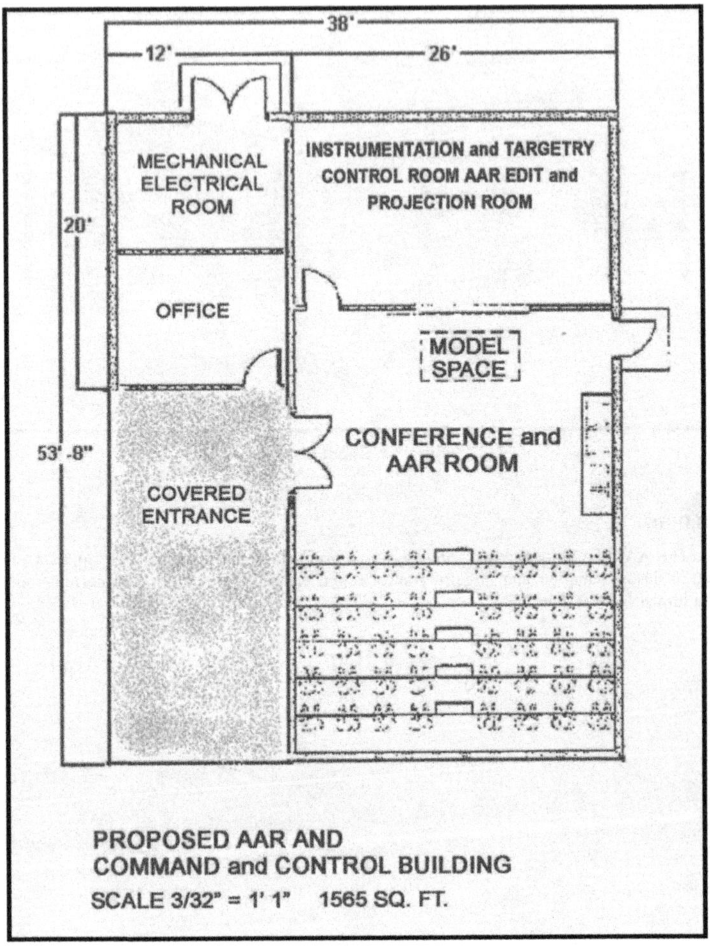

Figure 3-2. Command and control building and AAR room.

TRAINING AUDIENCE

3-3. The facility provides leaders the ability to train and evaluate the unit during a live-fire exercise. However, the unit's training requirements determine the size of the unit and the tasks to be trained at this facility. The unit develops its own organization and the steps for conducting the exercise based on its METL and training needs.

TRAINING INFORMATION

3-4. Tactical operations in an urban environment may range from combat missions to stability and support missions. Commanders should note that the Shoot House *is not* suited for the employment of live-fragmentation/concussion grenades, 40-mm HE/HEDP, antiarmor weapons, mortars, artillery, or air delivered munitions. Crew-served weapon systems, M2 and M1 tanks, and machine guns can be employed against supplemental targetry that is exterior to the Shoot House. The Shoot House has specially constructed areas that are used to conduct either mechanical or explosive breaching techniques. When conducting explosive breaching, units should consult with local Engineers for appropriate DES charges and consult Appendix A before conducting the training.

CONDUCT OF TRAINING

3-5. The unit develops realistic time frames for each operation or task. The times are based on the *crawl-walk-run* concept. The training event or STX can be arranged in any order as long as they follow a tactical scenario. *Assault a Building*, *Enter a Building and Clear a Room*, and *Clear a Building* are the primary collective tasks trained at this facility.

EXECUTION

3-6. Tables 3-1 through 3-5, pages 3-4 through 3-7, shows a sequence of events and times, and Figure 3-3 through Figure 3-6, pages 3-6 through 3-9, show example scenarios. These are only examples. The unit should develop training events based on its METL and training needs. However, the leader should do the following when conducting training at the live-fire Shoot House:

1. Identify the unit's critical missions (see commander's training guidance), and select those tasks that support specific tasks related to urban operations.
2. Arrange the related tasks in the order of their importance to mission accomplishment.
3. Assess the unit's current level of proficiency (strengths and weaknesses), and determine the degree of difficulty of each task. The training events and scenarios should start from the least difficult to the most difficult tasks. This gives the Soldiers confidence.
4. Identify the individual and leader tasks to be trained.
5. Conduct individual training first, because it is the first step in the building-block approach. It is also the basis for unit or collective task training. Mastery of individual skills is a precondition for effective training.
6. Set up training conditions based on unit proficiency. At first, keep the conditions basic, until Soldiers attain proficiency. Increase difficulty and realism as training progresses. Then, add limited visibility, CBRN factors, and live fires with service or short-range training ammunition (SRTA).
7. Understand that the Shoot House is a live-fire capable training facility. However, it can be used to train any level of proficiency, including blank, TES, SESAM, SRTA, and ball (no tracers). The leader assesses the unit's proficiency.

Table 3-1. Sequence of events and times for the platoon.

Element	*Event action*	*Estimated Time*
All 1st Squad 2d Squad 3rd Squad	1. Conduct precombat checks (for example; install, align, and troubleshoot MILES equipment; conduct inspection; draw equipment). 2. Support Operations. 3. Conduct Rehearsals. 4. Conduct Event 1 (Crawl).	2 Hours
3rd Squad 1st Squad 2d Squad	5. Support Operations. 6. Conduct Rehearsals. 7. Conduct Event 1 (Crawl).	2 Hours
2d Squad 3rd Squad 1st Squad	8. Support Operations. 9. Conduct Rehearsals. 10. Conduct Event 1 (Crawl).	2 Hours
1st Squad 2d Squad 3rd Squad	11. Support Operations. 12. Conduct Rehearsals. 13. Conduct Event 1 (Walk).	2 Hours
3rd Squad 1st Squad 2d Squad	14. Support Operations. 15. Conduct Rehearsals. 16. Conduct Event 1 (walk).	2 Hours
2d Squad 3rd Squad 1st Squad	17. Support Operations. 18. Conduct Rehearsals. 19. Conduct Event 1 (walk).	2 Hours
1st Squad 2d Squad 3rd Squad	20. Support Operations. 21. Conduct Rehearsals. 22. Conduct Event 1 (Run).	1 Hours
1st Squad 2d Squad 3rd Squad	23. Support Operations. 24. Conduct Rehearsals. 25. Conduct Event 1 (Run).	1 Hours
1st Squad 2d Squad	26. Support Operations. 27. Conduct Event 1 (Run).	1 Hours

Table 3-2. Event 1, Squad Sequence (*Crawl*).

Element	*Event Action*	*Estimated Time*
1 (CRAWL)	**Terminal Learning Objective** Familiarize the squad with how to enter and clear a room during the crawl phase of training **Tasks** *Enter and Clear a Room (Squad)* • *Perform Movement Technique* • *Select Hasty Firing Position* *Conduct a Breach (Mechanical)* • *Employ Hand Grenades* • *Engage Targets with M16A2/M4* • *Engage Targets with M249 Machine Gun*	2 Hours

NOTES: 1. Perform controlled pairs and target discrimination techniques (FM 3-22.9).

2. Include techniques (FM 3-06.11) and performance measures (FM 3-21.8 and DTMS).

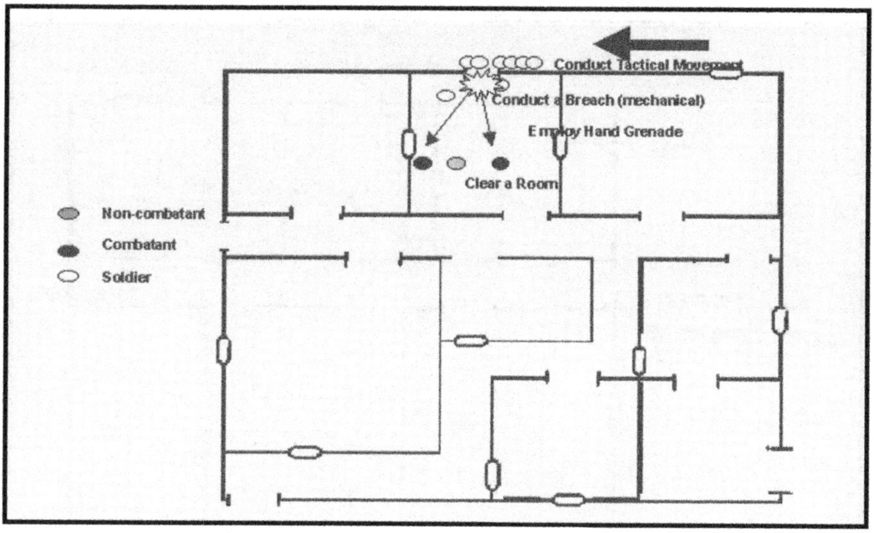

Figure 3-3. Event 1, *Squad Scenario (Crawl)*, force-on-targetry.

Table 3-3. Event 2, Squad Sequence (*Walk*).

Element	*Event Action*	*Estimated Time*
2 (WALK)	**Terminal Learning Objective** Familiarize the squad with how to enter and clear multiple rooms during the walk phase of training **Tasks** *Enter and Clear a Building (Squad)* • *Perform Movement Technique* • *Select Hasty Firing Position* *Conduct a Breach (Explosive DES)* • *Engage Targets with M16A2/M4* • *Engage Targets with M249 Machine Gun* *Conduct AAR*	2 Hours
NOTES: 1. Perform controlled pairs and target discrimination techniques (FM 3-22.9) 2. Include techniques (FM 3-06.11) and performance measures (FM 3-21.8 and DTMS).		

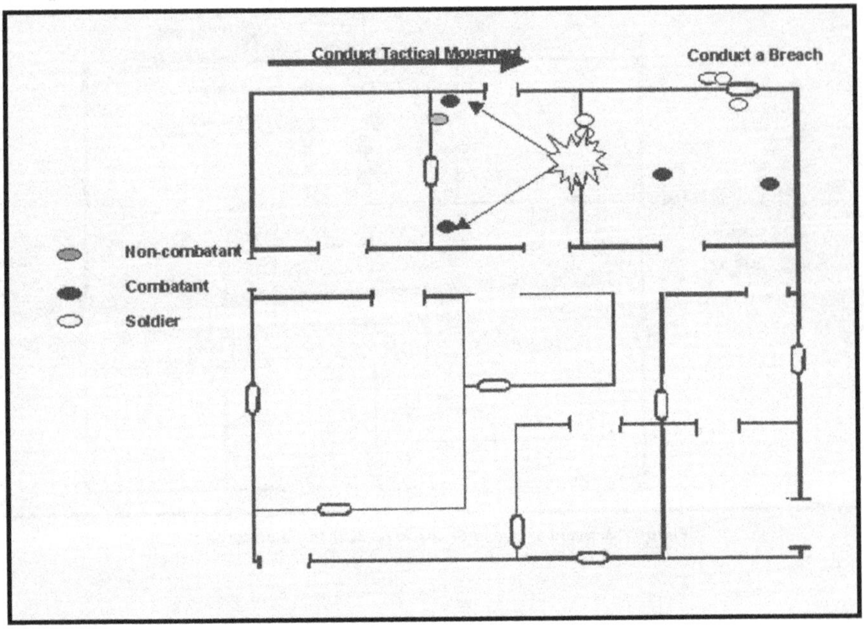

Figure 3-4. Event 2, *Squad Scenario (Walk)*, force-on-force.

Table 3-4. Event 3, Squad Sequence (*Run*).

Element	*Event Action*	*Estimated Time*
3 (RUN)	**Terminal Learning Objective** Refamiliarize squad with how to enter and clear multiple rooms during the run phase of training. **Tasks** *Enter and Clear a Building (Squad)* • *Perform Movement Technique* • *Select Hasty Firing Position* *Conduct a Breach (Mechanical)* • *Engage Targets with M16A2/M4* • *Engage Targets with M249 Machine Gun* *Clear a Hallway* • *Engage Targets with M16A2/M4* • *Engage Targets with M249 Machine Gun* *Conduct a Breach (explosive DES)* • *Engage Targets with M16A2/M4* • *Engage Targets with M249 Machine Gun* *Conduct AAR*	**1 Hour**

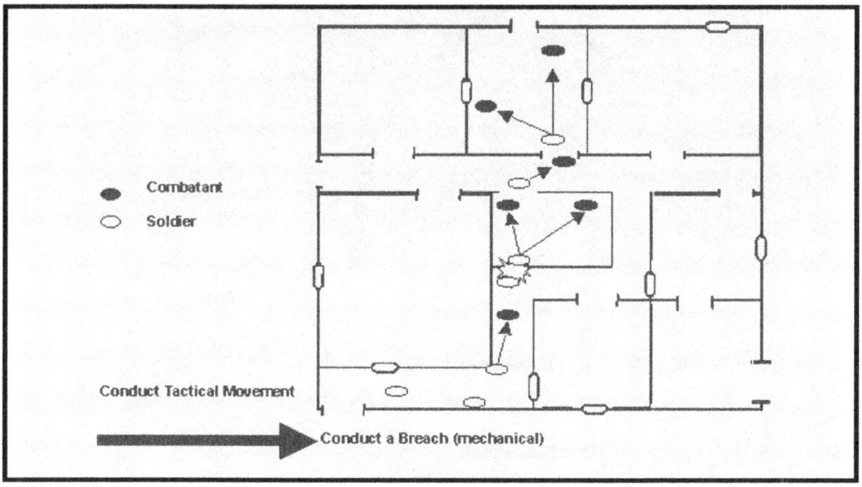

Figure 3-5. Event 3, *Squad Scenario (Run)*, force-on-targetry.

Table 3-5. Event 4, Platoon Evaluation Sequence (*Run*).

Element	Event Action	Estimated Time
4 (RUN)	*Occupy Assembly Area* *Develop and Communicate a Plan* *Prepare for Combat* *Perform Passage of Line* *Conduct Tactical Movement* *Break Contact* *Conduct Intermediate AAR* *Conduct Tactical Movement* *Perform Overwatch/Support by Fire* *Assault a Building* *Consolidate and Reorganize* *Conduct Final AAR*	1 Hour 2 Hours 30 min 1 Hour 5 min 20 min 1 Hours 1 Hours 2 Hours 1 Hours

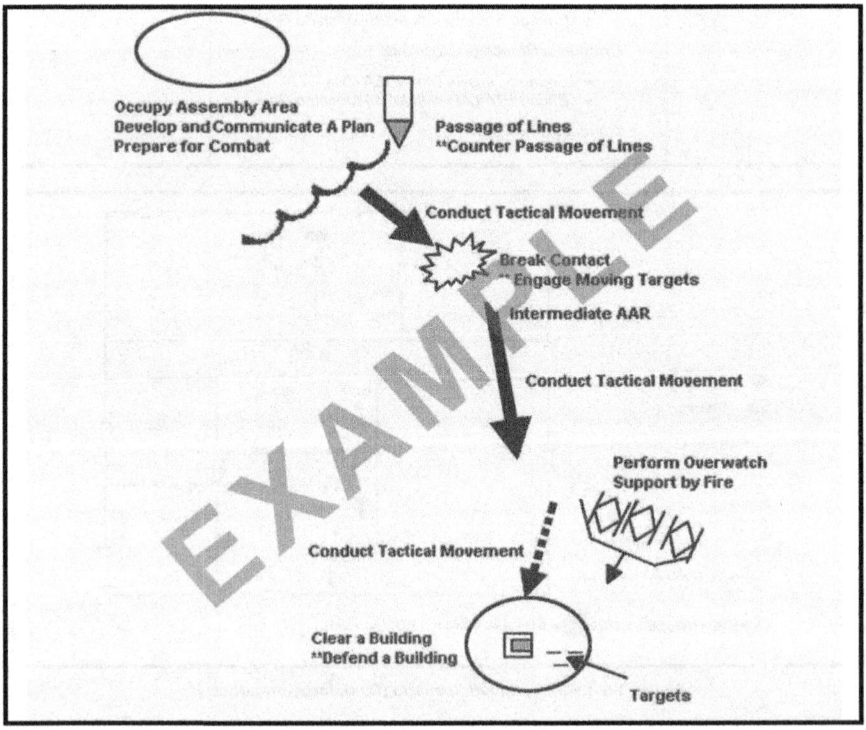

Figure 3-6. Example platoon tactical scenario (*Run*).

OPERATION ORDER

3-7. Figure 3-7 shows an example OPORD for the conduct of training.

```
                    ________________
                    (Classification)
                                        Copy ____ of ___ copies
                                        ____Company _____ Infantry
                                        (Unit Location)___________
                                        Date/time Group __________

(Message Reference Number)

OPORD: ______________
Reference Map, Series ____________________
Time Zone used Throughout the Order:_____________________

TASK ORGANIZATION:

______Platoon Platoon
______Platoon ____ 60-mm Section
______Antiarmor Section

1. SITUATION

   a. Enemy Forces. Annex B (Intelligence Overlay [prepared by
battalion S-2]).

      (1) The company is opposed by elements of the Red Company of the
Blue Brigade at about 60 percent strength. Within our zone, we have a
platoon-sized element occupying a defensive position in a building in
the vicinity of (______). The enemy has small arms weapons and may
call on company mortars for support.

      (2) Current indications are the enemy is regrouping to initiate
small-unit offensive actions in the area. His most likely course of
action is to move to the (direction) to link up with elements of the
parent unit.

   b. Friendly Forces.

      (1) (___) Infantry attacks at (______) hours to seize Objective
ALPHA (grid). The intent is to prevent the enemy element from moving
(direction) to link up with other enemy units.
      (2) Company (__) on our (left, right) makes the main attack at
(____) hours to seize Objective RED (grid) in order to establish
blocking positions to stop the enemy's movement to the (direction).

                    ________________
                    (Classification)
```

Figure 3-7. Example company OPORD.

(Classification)

2. **MISSION**

Company (__) attacks to seizes Objective BLUE (grid) at (date/time) hours in order to prevent the enemy from regrouping and engaging Company (__) (main effort) from Objective BLUE.

3. **EXECUTION**

a. **Concept of Operation**. Annex C (Operations Overlay). (This should include those steps listed in the task , Conduct Troop-Leading Procedures, paragraph 17. These should be written out when possible and not just placed on an overlay.)

(1) Maneuver. (___) Platoon establishes a defense in its zone, NLT (date/time) hours, to destroy small enemy elements moving or exfiltrating to the (direction), from the vicinity of Objective BLUE. This defense supports the attack on Objective BLUE by completing destruction of any enemy that may escape. (Evaluated) Platoon conducts the company main attack at (hours) to seize Objective BLUE to destroy the enemy or force them to withdraw into (___) Platoon's defense. (___) Platoon supports the attack of the (Evaluated) Platoon on Objective BLUE to suppress enemy fires and isolate the objective. On seizure of Objective BLUE, (Evaluated) Platoon and (___) Platoon will defend in sector to destroy the enemy elements withdrawing to avoid (___) Company's attack.

(2) **Fire support**.

(a) Priority of fires.

- Initially to (Evaluated) Platoon.
- Upon seizure of Objective BLUE, priority shifts to (Defensive) Platoon.

(b) Priority targets.

- 60-mm priority target initially AA4031.
- On initiation of assault on Objective BLUE, priority shifts to target AA4027 to isolate the objective.
- Upon seizure of Objective BLUE, priority shifts to target AA4025.

(Classification)

Figure 3-7. Example company OPORD (continued).

(Classification)

b. **Tasks to maneuver units.**

(1) (___) Platoon. Destroy enemy elements moving to the (direction), NLT (date/time).

(2) (Evaluated) Platoon. Consolidate and defend in sector to destroy enemy elements following the seizure of Objective BLUE.

(3) (___) Platoon. Be prepared to assault Objective BLUE, on order, consolidate and defend in sector to destroy enemy elements following the seizure of Objective BLUE.

c. **Tasks to Subordinate Units.**

(1) 60-mm Section. Initially in the vicinity of CP 7, priority of fires to the (Evaluated) Platoon. Move with the (Supporting) Platoon. On order, priority of fires to the (Defensive) Platoon.

(2) Antiarmor Section. Attached to the (Supporting) Platoon.

d. **Coordinating Instructions:**

(1) (Defensive) Platoon reports defense established and locations.
(2) Platoon reports the seizure of all objectives.
(3) Platoon reports the consolidation and establishment of the defense.
(4) Order of march from the assembly area to the objective: defense platoon, evaluated platoon, headquarters, support platoon, 60-mm mortar.

4. **SERVICE SUPPORT**

a. **General.** Company trains collocated with the battalion combat trains in the vicinity of (grid). Supply distribution at the company CP.

(Classification)

Figure 3-7. Example company OPORD (continued).

(Classification)

b. **Materiel and Services**.

(1) Class I: Ration cycle MRE-MRE-MRE.

(2) Class V: Pick up basic load at company CP, NLT (date/time).

(3) Medical: Company casualty collection point is __________.

(4) EPW: EPW collection point is ___________.

5. **COMMAND AND SIGNAL**

a. **Command**. Company command group initially follows the (Evaluated) Platoon.

b. **Signal**.

(1) Radio-listening silence is in effect until contact is made.

(2) SOI _____ is in effect.

JONES
Commanding

AnnexES:
B-Intelligence Overlay
C-Operations Overlay
D-Fire Support Overlay

(Classification)

Figure 3-7. Example company OPORD (continued).

TRAINING SUPPORT REQUIREMENTS

3-8. Minimum trainers and evaluators should be based on unit proficiency, training needs, and local safety requirements. Table 3-6 contains consolidated support requirements.

Table 3-6. Consolidated support requirements.

Ammunition	*Event 1*	*Event 2*	*Event 3*	*Event 4 (Optional)*
5.56-mm (per weapon) ball, blank, SRTA, or SESAMS	20 Blank	20 SESAMS	20 Ball/SRTA	20 Ball/SRTA
5.56-mm linked (per weapon)	30 Blank	30 SESAMS	30 Ball/SRTA	30 Ball/SRTA
Hand grenade body (practice)	0	--	4	4
Hand grenade fuzes (practice) per platoon	0	--	4	4
Hand grenade, smoke, per platoon	0	6	6	6
M84 hand grenade, stun, per platoon	0	4	4	0
TES (MILES Equipment)				
M16 (1 per weapon, 1 per OPFOR)	NO	YES	NO	NO
M249 (1 per weapon)	NO	YES	NO	NO
SAAF (1 per platoon)	NO	YES	NO	NO
Controller Gun (1 per OC)	NO	YES	NO	NO
Targetry	5	0	5	10
OPFOR	NO	6	NO	NO

TRAINING AND EVALUATION

3-9. The T&EOs are the foundation of the MTP and the collective training of the platoon. T&EOs are training objectives (task, conditions, and standards) for the collective tasks that support platoon critical wartime operations. T&EOs may be trained separately, in an STX, in an FTX, or in live-fire exercises. For collective live-fire standards, the trainer needs to refer to the applicable Infantry Gunnery Manual for the appropriate course of fire. Those standards and courses of fire need to be integrated into the training exercise. Table 3-8 provides a list of related collective and supporting individual tasks.

Table 3-7. Related collective and supporting individual tasks.

Task	FM 3-21.8	FM 3-06.11	STP 21-1-SMCT	STP 7-11B1-SM-TB	STP 21-24-SMCT
Conduct Consolidation and Reorganization Treat Casualties Evacuate Casualties Process Enemy Prisoners of War/Captured Materiel Defend MOUT Building Conduct Tactical Movement	X	X			
Supporting Individual Tasks					
Engage Targets with an M16-Series Rifle			071-311-2007		
Engage Targets with an M16-Series Rifle Using a Night Vision Sight AN/PVS-4			071-315-2007		
Engage Targets with an M249 Machine Gun				071-010-0006	
Employ Hand Grenades			071-325-4407		
Move Over, Through, or Around Obstacles (Except Minefields)			071-326-0503		
Neutralize Booby Traps					052-193-1013
Move as a Member of a Fire Team				071-326-0501	
Perform Movement Techniques during MOUT		X		071-326-0541	
Conduct the Maneuver of a Squad				071-326-5611	
Conduct an Explosive Breach				071-440-0030	
Execute a Mechanical Breach				071-440-0027	
Conduct a Breach				071-440-0029	
Operate Night Vision Goggles AN/PVS-7				071-710-0008	
Select Hasty Firing Positions during MOUT				071-326-0557	
Control Movement of a Fire Team				071-326-5605	

Chapter 4
Breach Facility

Urban operations require Soldiers to enter buildings through walls, doors, windows, and roofs. The Breach Facility is designed to train Soldiers using tasks and techniques on how to breach locked doors, windows, and create man-sized holes in walls. The facility can be used to train mechanical, ballistic, thermal, and explosive breaching.

PURPOSE

4-1. The Breach Facility (Figure 4-1) is used to train Soldiers semiannually on the technical aspects of breaching techniques. It is also used to train TTP and explosive techniques not trained on any other type range.

Figure 4-1. Breach facility.

DESCRIPTION

4-2. This facility has three stations:

STATION 1

4-3. This station is a facade that is 8 feet high and 88 feet wide with four sections of two doors for a total of eight doors. Soldiers can conduct mechanical, thermal, ballistic, and explosive breaching techniques on doors at this station. Half of the doors open from the exterior, and the other half of the doors open from the interior (Figure 4-2).

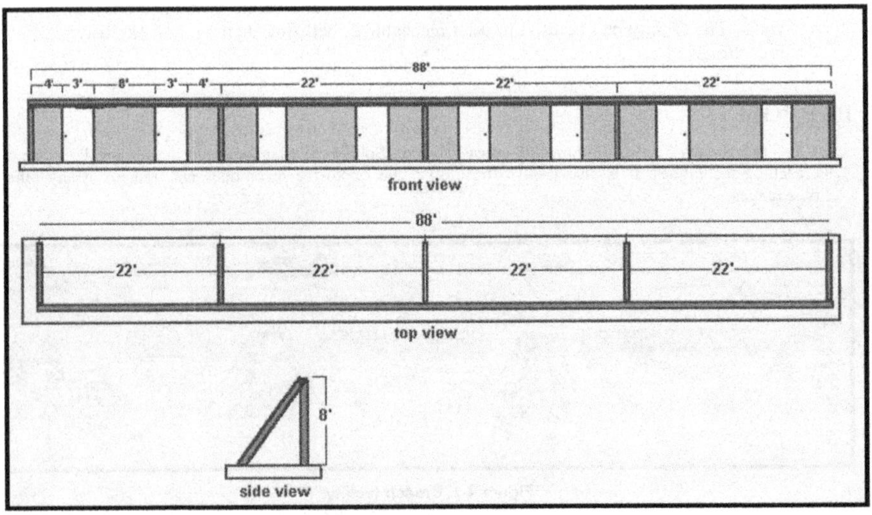

Figure 4-2. Station 1—Door breaching.

STATION 2

4-4. This station is a facade. It measures 10 feet high by 88 feet wide. It has four sections of two windows each, for a total of eight windows. Soldiers can conduct mechanical and explosive breaching techniques on windows at this station (Figure 4-3).

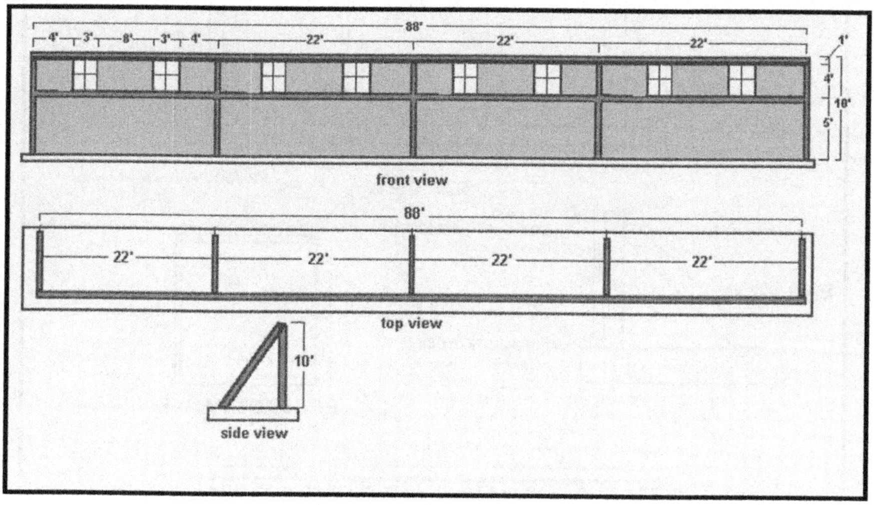

Figure 4-3. Station 2—Window breaching.

STATION 3

4-5. This station is a facade that looks like a wall of a building. It measures about 8 feet high by 28 feet wide. Precast panels will fit into each of the three sections. The panels measure 8 feet square and can be made of concrete, concrete blocks, stone, brick, wood or other material. The panels are designed for use with thermal and explosive breaching techniques (Figure 4-4).

TRAINING AUDIENCE

4-6. The facility provides Infantry, Engineer, MP, and Cavalry leaders with a facility to train and evaluate individual and leader breaching tasks. Unit training requirements determine the size of the unit and the tasks to be trained at this facility (Appendix B provides a unit training matrix). The individual tasks trained at the Breach Facility follow:

- 052-193-1003, *Prime Explosives Nonelectrically.*
- 071-440-0027, *Execute a Mechanical Breach.*
- 051-193-1055, *Construct a Nonelectric Initiating Assembly with Modernized Demolition Initiators (MDI).*
- 052-193-1312, *Construct Demolition Initiating Sets.*
- *Conduct a Thermal Breach* (Engineers only).

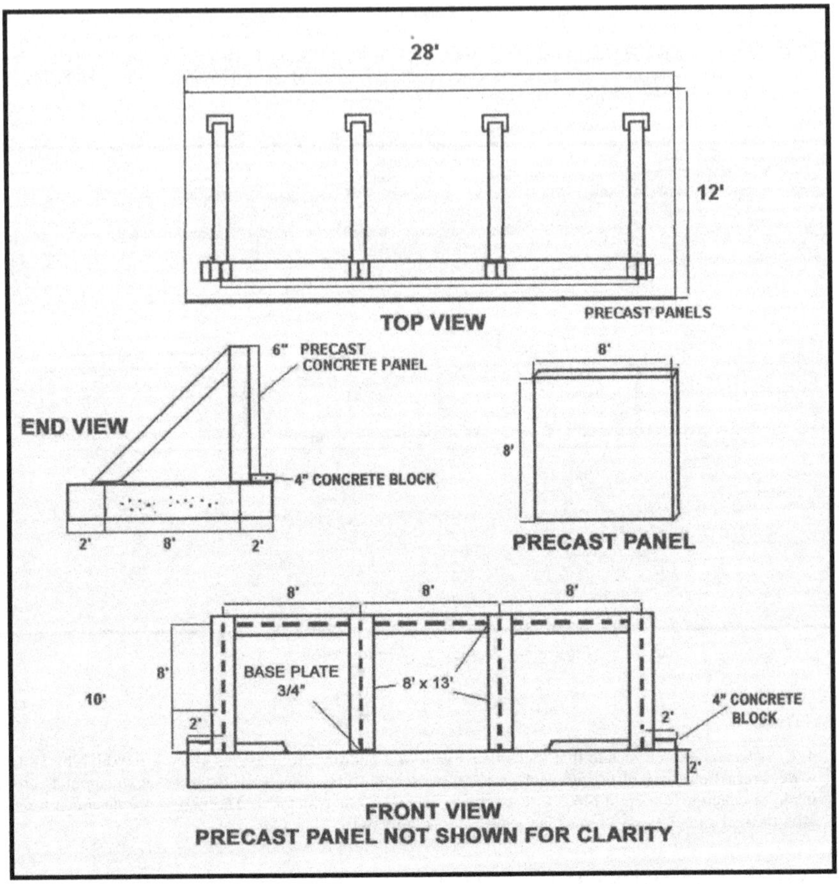

Figure 4-4. Station 3—Wall breaching.

TRAINING INFORMATION

4-7. Many types of organizations can train at this facility. Tactical operations in an urban environment may range from combat missions to stability and support actions. Commanders should note that this facility is an individual task and technique trainer. Training at the Breach Facility should be conducted prior to conducting collective training at the UAC, shoot house, and the CACTF. Trainers *must be certified* before conducting demolition training per local SOP or ACOM policy.

CONDUCT OF TRAINING

4-8. Trainers develop realistic time requirements for each task to be trained based on the *crawl-walk-run* concept and the size of the unit to be trained. The training event may be arranged in any order as long as it builds from simple to complex task and techniques. Table 4-1 contains an example sequence of events to assist units in developing their training plan. Leaders should do the following prior to conducting training. Before conducting ballistic breaching with automatic weapons, leaders should consult FM 3-06.11:

- Identify the unit's critical missions (see commander's training guidance) and select those tasks that support specific METL related to urban operations.
- Arrange the related tasks in the order of their importance to mission accomplishment.
- Assess the unit's current level of proficiency (strengths and weaknesses) and determine the degree of difficulty of each task. The training events should start from the less difficult tasks to the more difficult tasks. This instills confidence in the Soldiers conducting the training. Leaders should consult FM 3-06.11 for more information on explosive breach techniques.
- Identify the individual and leader tasks that are to be trained.
- Set up training conditions based on unit proficiency.

Table 4-1. Example platoon sequence of events.

Element	*Event Action*	*Estimated Time*
3d Sqd	Wall Breach: Explosive (crawl, walk, run)	3 hours
1st Sqd	Door Breach: Mechanical, ballistic, explosive (crawl, walk, run)	
2d Sqd	Window Breach: Mechanical, explosive (crawl, walk, run)	
2d Sqd	Wall Breach: Explosive (crawl, walk, run)	3 hours
3d Sqd	Door Breach: Mechanical, ballistic, explosive (crawl, walk, run)	
1st Sqd	Window Breach: Mechanical, explosive (crawl, walk, run)	
1st Sqd	Wall Breach: Explosive (crawl, walk, run)	3 hours NOTE: Walls, doors, and windows require replacement after each run/iteration to train to standards.
2d Sqd	Door Breach: Mechanical, ballistic, explosive (crawl, walk, run)	
3d Sqd	Window Breach: Mechanical, explosive (crawl, walk, run)	

TRAINING SUPPORT REQUIREMENTS

4-9. Table 4-2 shows minimum requirements based on three breach teams for each platoon-sized element. This facility is resource intensive, and Table 4-3 shows the total annual requirements per platoon-sized element.

Table 4-2. Consolidated minimum support requirements.

Ammunition/Equipment	*Crawl*	*Walk*	*Run*
Mechanical Breach: Prying , Striking, or Cutting Tool			
Door	(talk through)	1 per team (defeat one lock)	1 per team (defeat hinges)
Window	(talk through)	1 per team (defeat one lock)	1 per team (defeat hinges)
Ballistic Breach: Door			
Shotgun [1]	(talk through)	2 rds per weapon (defeat one lock)	5 rds per weapon (defeat hinges)
GREM (inert) [2]	(talk through)	1 per team – inert	1 per team - service
[1] If available to unit. [2] Grenade Rifle Entry Munition (GREM).			
Explosive Breach:			
Door: Det Cord (Flex Linear Charge) Nonelectric Firing System	(build)	(practice emplacement)	29 feet per team 1 per team
Wall: C4 (Satchel Charge) Nonelectric Firing System	(build)		4 blocks per team
		(practice emplacement)	23 feet per team 1 per team
Window: Det Cord (Flex Linear Charge) Nonelectric Firing System	(build)		29 feet per team 1 per team
		(practice emplacement)	
Thermal Breach Equipment Sappers (Only)	1	1	1

Table 4-3. Consolidated annual support requirements.

Ammunition/Equipment	*Total*
Doors	6
Windows	6
Wall Panels	6
Shotgun Rounds	42
GREM	6 inert/2 service
Det Cord (feet)	486
C4 (blocks)	24
Nonelectric Firing Systems	18

TRAINING AND EVALUATION

4-10. Individual training is the first step in the building-block approach and sets the basis for unit or collective training. Mastery of individual skills is an essential precondition for urban operations training and a good method for conducting individual training is through concurrent training. Concurrent training is designed to train groups of Soldiers on different tasks or techniques and these tasks or techniques may or may not be related. For example, a leader may subdivide the unit at the Breach Facility into breaching orders. One group of Soldiers may be conducting an explosive breach, while other groups are conducting mechanical and ballistic breaches. Ballistic breaching and explosive breaching does have a SDZ, an MSD, and ricochet considerations.

SAFETY CONSIDERATIONS

4-11. Safety requires effort on the part of everyone. It is not enough to teach safety rules; the rules must be practiced and enforced. Depending on the nature of the obstacle, the breacher may use mechanical, ballistic or explosive techniques. The following guidelines should be followed when training at the Breach Facility.

- Wear all appropriate personal protective equipment and follow appropriate guidelines:
  -- Safety glasses.
  -- Hearing protection.
  -- Body armor.
  -- Helmet.
  -- Gloves.
  -- Elbow and knee pads.
- Inspect equipment before use.
- Use appropriate tool.
- Use appropriate amount of explosive not to exceed four blocks of C4. Place the C4 on the wall only, not on the foundation, as this would cause irrepairable damage.
- Stay within the surface danger zone.
- Always use the minimum safe distance.
- Do not modify tool.
- Ensure proper use and maintenance of hydraulic and pneumatic tools; tools can burst under pressure when improperly used.

Chapter 5

Combined Arms Collective Training Facility

The home station CACTF is designed to conduct multiechelon, full-spectrum operations training up to battalion TF level. Example company and platoon STX lanes, and stability and support operations tasks, are provided to help units develop training plans. Leaders use training exercises to train, evaluate, and practice performance-oriented collective tasks. TEWT, STX, and FTX are the three types of exercises covered in this chapter. Each is designed to help the trainer develop urban operations training skills and increase unit proficiency in an urban environment. The urban operations collective training exercises were developed using the standard CACTF as described in this chapter and may be modified to fit available urban operations facilities.

PURPOSE

5-1. The leader or commander is provided with a CACTF to evaluate unit urban operations proficiency. The CACTF is intended to support blank fire, MILES/TES, SESAMS, STX, and FTX scenarios on a semiannual basis. The CACTF supports branch-specific lane training and combined arms training up to battalion level across the full spectrum of operations: offense, defense, stability, and support.

DESCRIPTION

5-2. The CACTF replicates an urban environment. The facility consists of 2.25 square kilometers of urban sprawl with 20 to 26 buildings, roads, alleys, parking areas, underground sewers, parks, athletic fields, and command and control building. The actual size and configuration of the CACTF depends on the local installation site requirements. The CACTF is designed to support heavy and light Infantry, Armor, artillery, and aviation positioning and maneuver. Table 5-1 shows the recommended types of buildings and features in the CACTF.

Table 5-1. CACTF characteristics.

20 to 26 Buildings (1.5 km by 1.5 km)	
Tunnel and sewer system Shantytown One 3-story building Three 2-story buildings Industrial area Electricity and potable water City dump	Props and furniture Targetry Audio and image capture with EDIT/REPLAY capability Control building with AAR facility Breachable walls

5-3. The buildings have one, two, or three stories (some with basements), sloped and flat roofs. CACTF building variations pose different tactical and technical training challenges. Figure 5-1 shows an example CACTF containing 24 buildings with the following characteristics:

- The hotel is a dominating structure typical of a central business district. The hotel has an elevator shaft, fire escapes, and a large first-floor lobby.
- The two-story office building contains a series of identical rooms.
- The school has a long, central corridor, large windows, and a single large room.
- The townhouse has multiple sections with a common attic.
- The service station accommodates combat vehicles in its service bay.

5-4. The buildings have other features that enhance the CACTF's training value.

- Constructed mouse-holes permit movement between rooms and floors.
- Loopholes in roofs and outside walls allow observation and fields of fire.
- Roof hatchways lead to the top floors of multiple story buildings.

5-5. Rooms vary in size, and in number of doorways and windows. These variations require Soldiers to identify rooms that provide suitable indoor fighting positions for TOW, Javelin, Dragon, and AT4 antitank weapon systems.

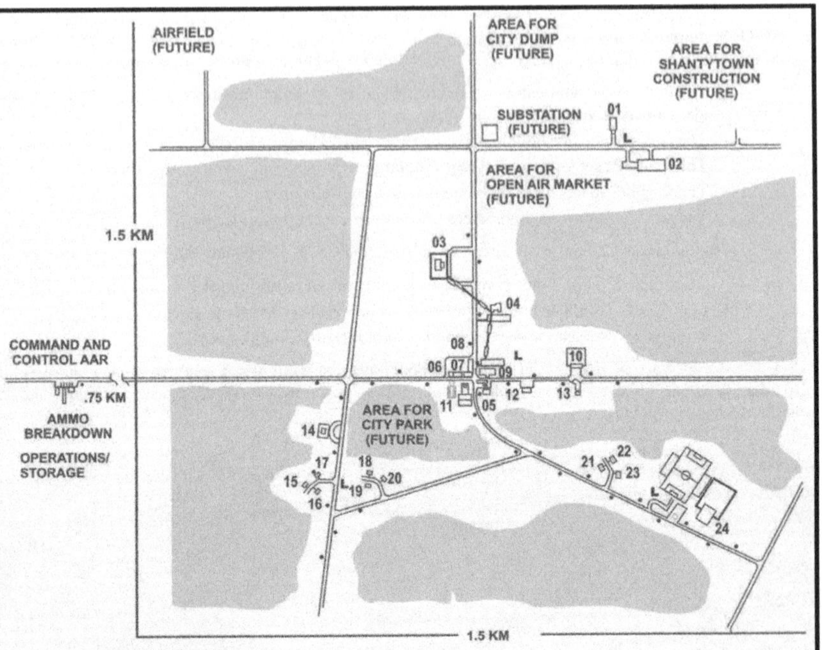

BLDG #	SIZE AND TYPE
01	1 STORY WAREHOUSE
02	1 STORY WAREHOUSE
03	3 STORY OFFICE (EMBASSY)
04	2 STORY OFFICE
05	1 STORY SERVICE STATION
06	1 STORY BUSINESS
07	1 STORY BUSINESS
08	1 STORY BUSINESS
09	3 STORY HOTEL
10	1 STORY POLICE STATION/JAIL
11	CHURCH AND CEMETARY
12	1 STORY BUSINESS
13	1 STORY BANK
14	2 STORY TOWNHOUSE
15	1 STORY RESIDENT
16	1 STORY RESIDENT
17	1 STORY RESIDENT
18	1 STORY RESIDENT
19	1 STORY RESIDENT
20	1 STORY RESIDENT
21	1 STORY RESIDENT
22	1 STORY RESIDENT
23	1 STORY RESIDENT
24	2 STORY SCHOOL

LEGEND

- FENCE WALL
- WALL
- TUNNEL SYSTEM
- TREE LINE
- UTILITY POLE
- LATRINE

SCALE: 1" = 500'

Figure 5-1. Example CACTF schematic.

5-6. Buildings are arranged in a realistic urban pattern. The CACTF represents an area consisting mainly of residential, commercial, public institutional, and light industrial buildings. Buildings and streets are arrayed with sufficient dispersion to allow up to nine platoon-sized STX sectors (Figure 5-2) or six company/team STX sectors (Figure 5-3).

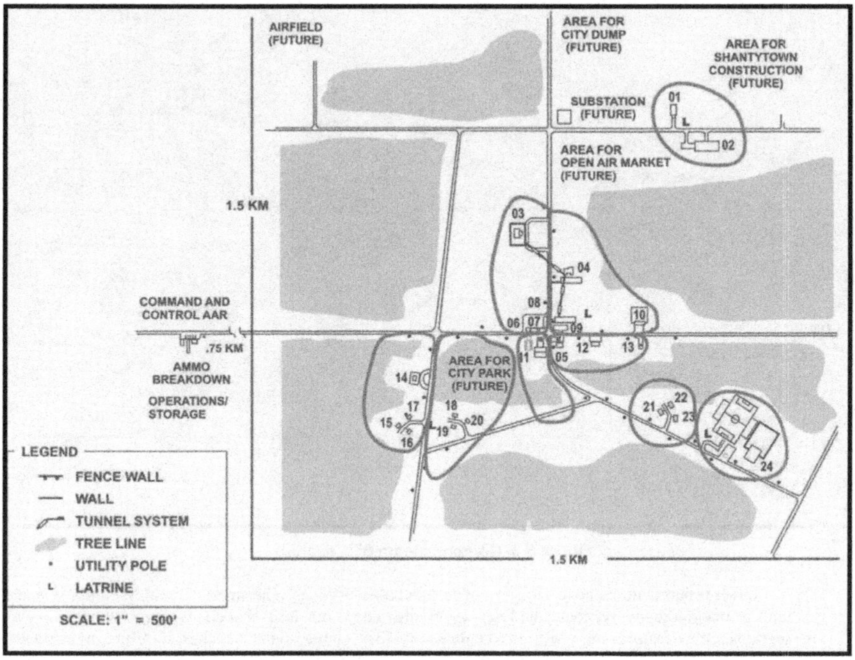

Figure 5-2. Platoon ODSS STX sectors.

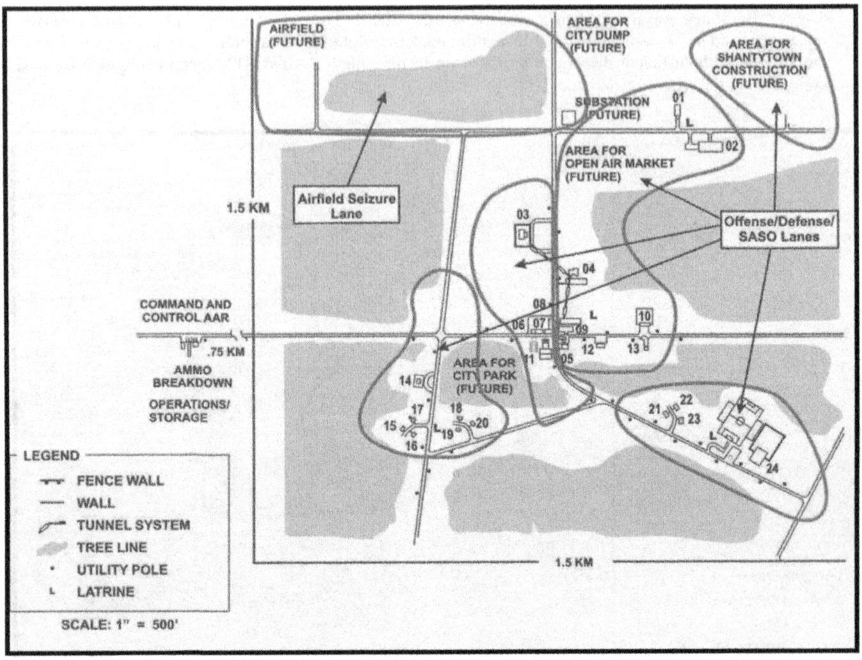

Figure 5-3. Company/team STX sectors.

5-7. Other urban features pose a variety of terrain considerations. The street network includes one and two-lane primary, secondary, local, and service roads. They join in T-shaped, L-shaped, and four-way intersections. Other features may be used to enhance realism such as signs, benches, furniture, newsstands, utility poles, street lights, vegetation, simulated rubble, live civilians on the battlefield, role players, and OPFOR. Open areas next to the CACTF can be used for maneuvering and staging areas. The specific details and layout of the CACTF may vary from one installation to another.

> **CAUTION**
>
> Trainers must ensure that fighting position reinforcement does not exceed the weight limits of the buildings in order to avoid structural damage and safety hazards to the Soldiers.

TACTICAL EXERCISE WITHOUT TROOPS

5-8. A terrain walk and a TEWT may be conducted in a local downtown area to prepare leaders and to take advantage of the available time in the CACTF. Terrain walks and TEWTs are conducted to teach leaders the tactical value of urban terrain. Appendix A provides training tips and additional information regarding the conduct of a TEWT.

COLLECTIVE TRAINING EXERCISES

5-9. The following exercises were developed IAW the guidance in FM 7-0, FM 7-1, FM 25-4, and TC 7-98-1. These publications include offensive and defensive scenarios for combat in urban areas, specifically, two platoon STXs and two company FTXs. Each can be conducted with blank ammunition; with MILES/TES; with 9-mm or 5.56-mm SESAMS for short-range precision target engagements (less than 50 meters); or force on force. Each was chosen to help the commander develop, sustain, and evaluate his unit's mission proficiency:

- Infantry Rifle Platoon Attack STX
- Mechanized Infantry Platoon Defend STX
- Infantry Rifle Company Attack FTX
- Mechanized Infantry Company Team Defend FTX

INFANTRY RIFLE PLATOON STX ATTACK

1. **Objective.**

This example STX trains collective, leader, and individual tasks for the platoon operation, Attack in a Built-Up Area.

2. **Interface.**

This STX is supported by two warrior drills, *React to Contact* and *Enter and Clear a Building*.

3. **Training.**

a. *Guidance.* FM 7-0 and FM 7-1 provide training guidance, tips, and enhancers for planning and executing STXs. The trainer should review the individual, leader, and collective tasks to be performed during the STX to determine the ones that require initial or refresher training.

b. *General Situation.*

(1) The exercise scenario follows: Contact with the enemy has been reestablished. Initial reports indicate he is at 65 to 70 percent strength, and that he has not received reinforcements. His defensive positions are in the CACTF and are not well established. He has indirect fire and CAS; he has already used chemicals, and will probably do so again. The platoon receives an order to prevent the enemy from establishing a heavily fortified defense in the CACTF. Such a defense could prevent friendly forces from continuing offensive operations.

(2) This exercise begins when the platoon leader receives the company FRAGO, and it ends after the platoon consolidates and reorganizes on the objective. An AAR should be held after completing the task, *React to Contact*, and at the completion of the exercise. If necessary, portions of the exercise should be repeated until the platoon performs them to standard. Figure 5-4 provides a graphic scenario of task performance for the STX *Attack*. Table 5-2 provides a recommended sequence for performance of T&EOs, and the time allotted for each portion of the STX.

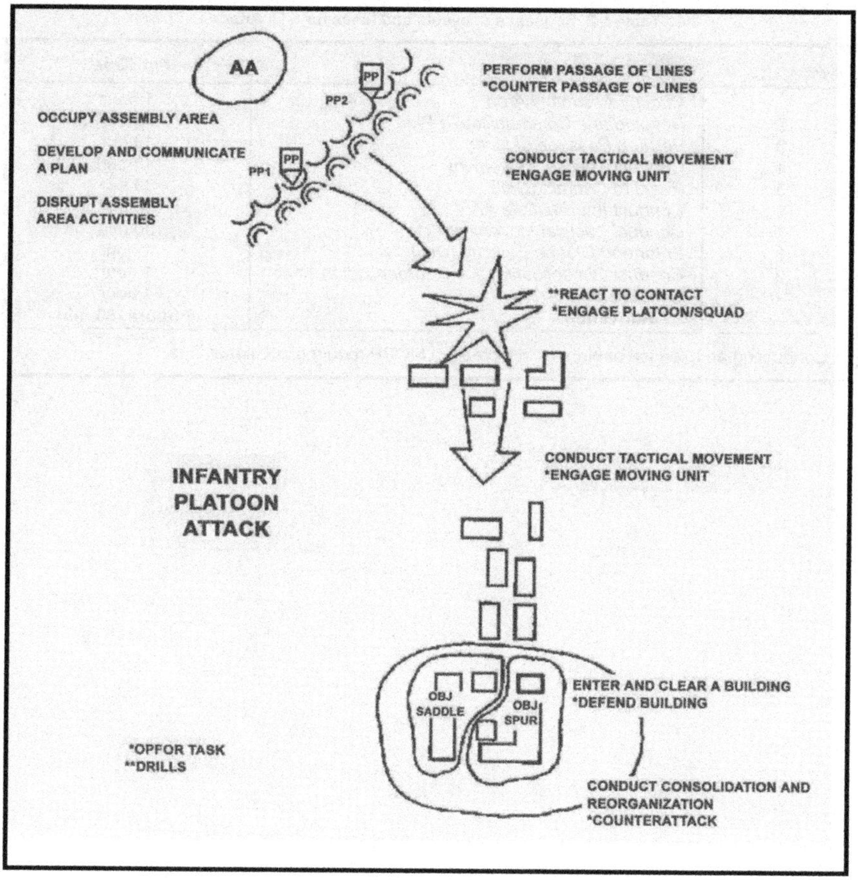

Figure 5-4. Attack STX scenario.

Table 5-2. Sequence of events and times for STX Attack.

Event	*Action*	*Estimated Time*
1	*Occupy Assembly Area*	1 hour
2	*Develop and Communicate a Plan*	1 hour
3	*Perform Passage of Lines*	1 hour
4	*Conduct Tactical Movement*	1 hour
5	*React to Contact (Drill)*	30 min
6	*Conduct Intermediate AAR*	30 min
7	*Conduct Tactical Movement*	30 min
8	*Enter and Clear a Building (Drill)*	1 hour
9	*Conduct Consolidation and Reorganization*	1 hour
10	*Conduct Final AAR*	1 hour
	* **Total Time**	**8 hours, 30 min**
* Conducting an exercise at night or in increased MOPP requires additional time.		

c. *Special Situation.* The platoon is part of a company in a secure assembly area when a FRAGO to attack is received (Figure 5-5).

1. **SITUATION.** The company is opposed by one platoon.

a. *Enemy Forces.* The enemy is at 65 to 70 percent strength. He is preparing a defense in CACTF (**MOUT site**) to prevent friendly forces from continuing to attack along the CACTF avenue of approach. He is expected to use artillery-delivered nonpersistent nerve agents in the defense of CACTF.

b. *Friendly Forces.* (Battalion designation) attacks (**date-time group**) to destroy enemy forces at OBJ COWBOY (**CACTF**) in order to pass follow-on forces, allowing friendly forces to continue the attack.

2. **MISSION.** (___) Company seizes OBJ HORSE (CACTF) vicinity (**grid**) NLT (**date-time group**) to prevent the enemy from establishing a heavily fortified defense.

3. **EXECUTION.**

a. *Concept of the Operation.* (See overlay.). The company moves in a wedge to Assault Position 1, assault CACTF to seize OBJ HORSE, consolidate, and reorganize.

(1) Maneuver. (**Evaluated**) Platoon will lead the company and be the main effort in the attack. They seize OBJ SADDLE vicinity (**grid**) to allow passage of friendly elements through CACTF. (___) Platoon moves on the company's left and seize OBJ SPUR vicinity (**grid**) to protect the left flank of the main effort. (___) Platoon moves on the company's right and establish a support-by-fire position located vicinity (**grid**).

(2) Fire support. fires are used to obscure movement to OBJ HORSE and to suppress the enemy during the assault. Priority of indirect fires is to (*evaluated*) Platoon, which is allocated one priority target (**120-mm mortar**).

Figure 5-5. Example FRAGO for STX Attack.

```
 b. Tasks to Maneuver Unit.
   (1)  (Evaluated) Platoon.
     (a)  Use Passage Point 1.
     (b)  Orient defense from TRP 2 to TRP 3.
     (c)  Be prepared to employ a wire-mined obstacle vicinity TRP 5
during consolidation.
   (2)  (___) Platoon.
     (a)  Use Passage Point 2.
     (b)  Orient defense from TRP 1 to TRP 2.
   (3)  (___) Platoon
     (a)  Use Passage Point 2.
     (b)  Orient defense from TRP 3 to TRP 4.
 c. Coordinating Instructions.
   (1)  Company RP is located (grid).
   (2)  Company linkup is (grid).
   (3)  MOPP1 is in effect (date and time).
   (4)  Selective rubbling is authorized. Destruction of an area
larger than one city block requires brigade commander approval.
```

Figure 5-5. Example FRAGO for STX Attack (continued).

4. **Support Requirements.**

a. *Minimum Trainers/Evaluators*: This exercise can be conducted with the company commander or platoon leader acting as the trainer or primary evaluator. At least one more controller or evaluator is required with the OPFOR. Other platoons being trained or evaluated will increase support personnel requirements.

b. *Vehicles/Communications*: All vehicles organic to the platoon should be included. The OPFOR should also have some combat vehicles.

c. *Opposing Force*:. The OPFOR should consist of at least a reinforced squad.

d. *Maneuver Area*: A training area at least 4 by 15 km is needed for cross-country movement and infiltration, and the CACTF is the ideal objective area. The terrain should offer multiple covered and concealed approaches to the objective.

5. **Evaluation.**

Table 5-3 lists the T&EOs from DTMS used to evaluate this STX.

Table 5-3. T&EOs used to evaluate STX Attack.

Task/Function	*Task Number*
Perform Passage of Lines	7-3-1125
Conduct Tactical Movement	7-2-1342
Occupy Assembly Area	7-3-1136
Develop and Communicate a Plan	7-3-1605
Conduct Consolidation and Reorganization (Company/Platoon)	7-2-5027

MECHANIZED INFANTRY PLATOON STX DEFEND

1. **Objective**. This example STX trains collective, leader, and individual tasks for the platoon operation, *Defend Urban Area/Building*.

2. **Interface.** This STX is supported by two warrior drills: *React to Contact* and *Enter and Clear a Building*.

3. **Training.**

a. *Guidance*. FM 7-0 and FM 7-1 provide training guidance, tips, and enhancers for planning and executing STXs. The trainer should review the individual, leader, and collective tasks to be performed during the STX to determine which require initial or refresher training.

b. *General Situation*.

(1) The exercise scenario is as follows: Latest INTSUMs indicate that enemy forces are expected to use a high-speed avenue of approach to counterattack. The approach is dominated by a small urban area containing strongly constructed buildings. The enemy has the capability for indirect fire, CAS, and chemicals. A defense in and around the town is necessary to deny the enemy the use of the avenue of approach. The platoon, as part of a larger force, is ordered to occupy and prepare defensive positions that overlook the avenue of approach. Indirect fire is available.

(2) Conduct of the exercise begins when the platoon receives an OPORD to move to establish a platoon defensive position. An AAR should be conducted after Soldiers enter a building and clear a room, and a final AAR should be conducted once all evaluation notes are compiled. If necessary, portions of the exercise should be repeated until the platoon's performance is satisfactory.

(3) When the platoon receives a warning order to prepare to move to establish a defense, the platoon is part of a company, which is part of a task force, and is located in an assembly area. Figure 5-6 provides a graphic scenario of task performance for the Defend STX. Table 5-4 provides a recommended sequence for performance of T&EOs and the time allotted for each portion of the STX.

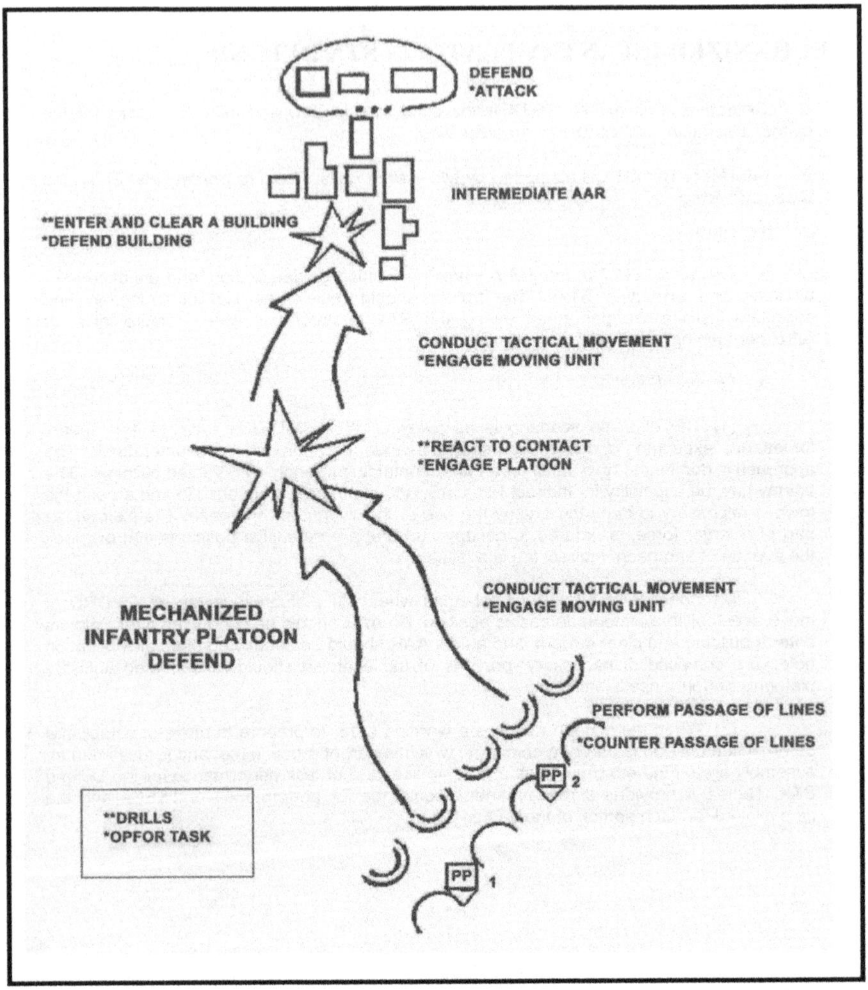

Figure 5-6. Defend STX scenario.

Table 5-4. Sequence of events and times for STX Defend.

Event	*Action*	*Estimated Time*
1	*Perform Passage of Lines*	1 hour
2	*Conduct Tactical Movement*	30 min
3	*React to Contact*	1 hours
4	*Conduct Tactical Movement*	30 min
5	*Enter and Clear a Building (Drill)*	1 hour
6	*Intermediate AAR*	30 min
7	*Defend MOUT/Building*	6 hours
8	*Conduct Final AAR*	1 hour
	Total Time	**11 hr 30 min**
*Conducting an exercise at night or in increased MOPP requires additional time.		

c. *Special Situation.* Thirty minutes after receiving the warning order, the platoon receives the company OPORD (Figure 5-7).

____
(Classification)

Copy no ___ of ___ copies
___ Battalion, ___ Infantry
Unit location or place of issue
Date-time group (with time zone)
Message reference number

OPERATION ORDER NO. _____
References: Map, series no. ___, sheet no. ___, edition ___
Time Zone Used Throughout the Order: ____
TASK ORGANIZATION:
___ **Platoon,** ___ **Platoon(-),** ___ **Platoon, Company Control**

Company HQ
1/1/C-52d Engr
1/2 ___,___
(Reserve)
Company Trains

1. **SITUATION.**

a. **Enemy Forces.**

(1) *Composition, Distribution, and Strength.* We are opposed by elements of the Liberation Guards Militia, Red Brigade. Several platoon-sized outposts are located within the battalion sector and consist of about 30 to 35 Soldiers per platoon. They are equipped with an assortment of Soviet small arms, including AK-47s, RPKs, RPGs, and various wheeled and tracked vehicles including some BMPs. Enemy outposts are at 60 to 70 percent strength and have indirect fire assets available.

(2) *Capabilities.* The enemy units are well entrenched in their defensive positions. The strongpoints have dedicated fire support assets and are capable of using nonpersistent nerve agents to prevent coordinated attacks. No enemy air threat exists but some heavy armored vehicles have been sighted in the area.

(3) *Probable Course of Action.* Intelligence indicates the enemy strongpoints will continue to fortify their positions until relief arrives within 24 hours in the form of a main enemy counterattack.

b. **Friendly Forces.** (__) TF defends in sector NLT (______) along PL (____) to deny enemy forces use of avenue of approach (____). (__) TF must be prepared to continue movement forward of PL (____), on order.

(1) Missions of units on left and right are (as required).

____
(Classification)

Figure 5-7. Example OPORD for STX Defend.

(Classification)

(1) (__) TF defends in sector NLT (______) from (grid) to (grid) to (grid) to contain any penetrations of the brigade sector.

(1) (___) Battery, (___) Battalion provides DS to 1st Brigade.

c. **Attachments and Detachments**. 1/1/C-52d Engineers attached.

2. MISSION. (___) Company blocks enemy forces NLT (date-time group) to prevent them from using high-speed avenues of approach through (name of MOUT site), and to prevent the enemy from counterattacking into the battalion sector.

3. EXECUTION.

a. **Concept of the Operation**. (See overlay.) Company prevents the enemy from using the high-speed avenues of approach through (MOUT site) by moving to the site and establishing a blocking position, which overwatches the platoon BPs covering the main avenues. We will block in depth by placing two platoons forward and one back, and maintaining a reserve to block any penetration. We must ensure our sectors of fire are placed to mass fire at the decisive point.

(1) *Maneuver*. (Evaluated) Platoon acts as the main effort, blocking from BP (__), vicinity (grid), to prevent enemy penetration of PL (____). (___) Platoon occupies BP (__), vicinity (grid), to block enemy movement into the main effort's left flank. (___) Platoon occupies BP (__) to prevent the enemy from making a coordinated attack on the main effort.

(2) *Fire Support*. The purpose of fires is to disrupt the enemy when his counterattack reaches PL (____). Priority of indirect fires is to (*evaluated*) Platoon, which is allocated two 155-mm priority targets, one of which may be used as an FPF.

(3) *Intelligence*. Priority of collection is to identify and locate enemy combat reconnaissance patrols and the enemy main effort.

(4) *Engineering*. During movement to the battle position, obstacles are used to ensure mobility. During blocking actions, the purpose of obstacles is to prevent the enemy from advancing on larger streets and to canalize him into more restrictive terrain. Priority of effort is to countermobility. Priority of support is to (*evaluated*) Platoon.

b. **Tasks to Maneuver Units**:

(1) *(Evaluated) Platoon*:

(a) Conduct passage of lines using PP 1.

(b) Lead the company movement after crossing the LD. Be prepared to react to enemy contact when (MOUT site) is reached.

(c) Enter and clear building(s) (number determined by commander).

(Classification)

Figure 5-7. Example OPORD for STX Defend (continued).

(Classification)

(d) Occupy BP (__) and orient on TRPs 2 and 3.

(e) Establish an OP vicinity (grid) to observe main avenue of approach.

(2) *(___) Platoon:*

(a) Conduct passage of lines using PP 2.

(b) Move second in order of movement.

(c) Enter and clear building(s) (number determined by commander).

(d) Occupy BP (__) and orient on TRP 1.

(3) *(___) Platoon:*

(a) Conduct passage of lines using PP 3.

(b) Move last in order of movement.

(c) Enter and clear building(s) (number determined by commander).

(d) Occupy BP (__) and orient on TRPs 4 and 5.

(e) Establish an OP vicinity (grid).

c. **Tasks to Subordinate Units**: Engineer squad travels with the company headquarters behind (*evaluated*) Platoon and supports the main effort's emplacement of obstacles.

d. **Coordinating Instructions.**

(1) Timings.

(a) Brief back ________.

(b) Company rehearsal ________.

(c) SP time ________.

(d) In position NLT (time) ________.

(2) Report all enemy contact.

(3) Report crossing all phase lines.

(4) MOPP status ____; enemy chemical attack is expected.

(5) Air defense status is YELLOW, WEAPONS HOLD.

(6) Leaders briefback the movement and defense plan to the commander at the company CP at (date-time group).

(7) Destruction of subterranean approaches requires brigade commander's approval.

(8) Selective rubbling is authorized. Destruction of more than one building requires commander's approval.

(9) Marking of rooms and buildings is IAW CO SOP.

(Classification)

Figure 5-7. Example OPORD for STX Defend (continued).

(Classification)

(10) Rules of Engagement follow:
(a) Minimize collateral damage to structures.
(b) Civilians have been evacuated. Evacuate to company CP any other civilians encountered.

4. SERVICE SUPPORT. IAW the battalion SOP.

a. **Miscellaneous.**
(1) Battalion trains located vicinity (grid).
(2) Company trains located vicinity (grid).

b. **Materiel and Services.**
(1) *Supply.*
(a) Class I. Ration Cycle M-M-M. Platoons must maintain a three-day supply in each BP.
(b) Class II. Submit requirements for special equipment ASAP. Coordinate with 1SG.
(c) Class III. Refuel all vehicles before (date-time group).
(d) Class IV. Coordinate with S-4 for delivery.
(e) Class V.
- Platoons cache one additional basic load in BP.
- Platoons submit request for additional demolitions NLT (date-time group).

(2) *Services.* Battalion decontamination site is located vicinity (grid).
(3) *Maintenance.*
(a) Priority to (*evaluated*) Platoon M2 BFVs.
(b) UMCP located vicinity (grid).

c. **Medical Evacuation and Hospitalization.**
(1) Battalion aid station located vicinity (grid).
(2) Company CCP located with company CP.

d. **Personnel.**
(1) No replacements for next 72 to 96 hours.
(2) Brigade mortuary affairs point located in BSA.
(3) EPW collection sites located at (grid) for battalion and (grid) for company.

(Classification)

Figure 5-7. Example OPORD for STX Defend (continued).

(Classification)

5. COMMAND AND SIGNAL.

a. **Command.** Initially, company CP will be traveling center of company wedge. Location in BP to be determined.

b. **Signal.**
(1) Current SOI in effect.
(2) Primary means of communication in BP is wire, FM, messenger, in that order.
(3) Radio listening silence is in effect until after crossing PL (____) or until contact is made.

(Classification)

Figure 5-7. Example OPORD for STX Defend (continued).

4. **Support Requirements**.

a. *Minimum Trainers/Evaluators*: This exercise is conducted by one company grade officer who is the trainer and primary evaluator.

b. *Vehicles/Communications*: All vehicles organic to the platoon should be included.

c. *Opposing Force*:. The OPFOR should be platoon strength.

d. *Maneuver Area*: A training area at least 2 by 10 km, with hasty fighting positions, is desirable for cross-country movement. The CACTF serves as the objective area. The terrain, ideally, should offer multiple covered and concealed routes. Using terrain that limits the leader to a geographic or predictable route prevents evaluation of the unit's ability to conduct a terrain analysis and select covered and concealed routes.

e. *Barrier Materials*. Sufficient Class IV (sandbags, lumber, and so on) must be made available to fortify positions, IAW Chapter 3, FM 3-06.11.

5. **Evaluation**.
Table 5-5 lists the T&EOs from DTMS used to evaluate this STX, and Table 5-6 lists the supporting tasks.

Table 5-5. T&EOs used to evaluate STX Defend.

Task/Function	*Task Number*
Defend MOUT/Building	7-3-1118
Perform Passage of Lines	7-3-1125
Conduct Tactical Movement (Co/Plt)	7-2-1342

INFANTRY RIFLE COMPANY FTX ATTACK

Table 5-6. Supporting tasks for company FTX.

STX C-II-1	*STX C-II-2*	*STX C-II-3*
Conduct an Infiltration (Co/Plt) Perform Linkup Execute an Assault (MOUT) Conduct Consolidation and Reorganization (Co/Plt)	Perform Cordon and Search Take Action on Contact Conduct an Infiltration (Co/Plt) Perform Linkup Perform Passage of Lines Occupy Assembly Area	Disarm Belligerents Handle Captured Insurgents or Belligerents Apprehend/Detain Noncombatants

1. **Objective**. This FTX is designed to train the Infantry rifle company in conducting offensive operations. This exercise provides practice for the company commander, platoon leaders, squad leaders, and company headquarters personnel in planning, coordinating, and controlling combat operations.

2. **Interface**. This FTX is supported by STX C-II-1, C-II-2, and C-II-3 (Table 5-6). It is also supported by two warrior drills, *React to Contact* and *Break Contact.*

3. **Training**.

a. *Guidance*. FM 7-0 and FM 7-1 provide training guidance, tips, and enhancers for planning and executing STXs. The trainer should review the individual, leader, and collective tasks to be performed during the STX to determine which require initial or refresher training.

b. *General Situation.*

(1) The battalion is ordered to conduct offensive operations in support of the brigade mission. The company is ordered to infiltrate enemy lines and to destroy an insurgent operations base located in the CACTF. The company must be prepared to remain in the CACTF to conduct cordon and search missions.

(2) This exercise should be conducted under various conditions.

(3) When the company receives the battalion warning order the planning process for the attack mission begins. The exercise ends when the company completes a passage of lines, occupies an assembly area, and completes preparations to continue its mission. Figure 5-8 shows a graphic scenario of task performance for the FTX Attack. Table 5-7 provides a recommended sequence for performance of T&EOs and the time allotted for each portion of the STX, including AARs.

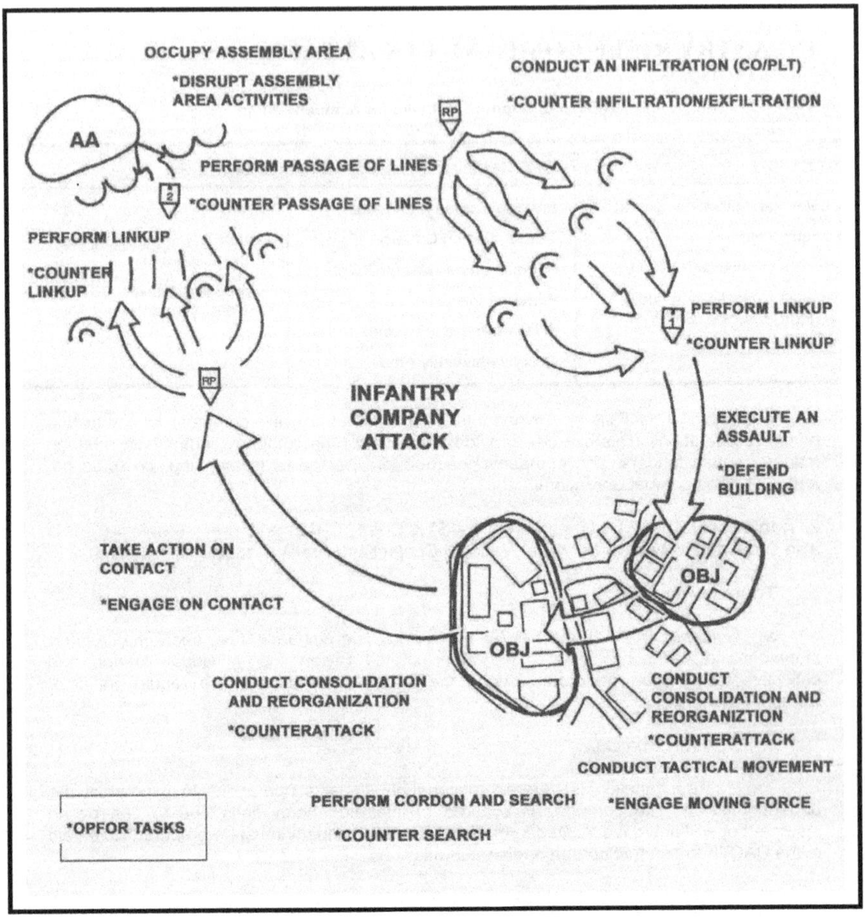

Figure 5-8. Attack FTX scenario.

Table 5-7. Sequence of events and times for FTX Attack.

Event	*Action*	*Estimated Time*
START STX C-II-1		
1	*Issue Company OPORD*	4 hours
2	*Conduct an Infiltration (Co/Plt)*	6 hours
3	*Perform Linkup*	3 hours
4	*Execute an Assault (MOUT) AAR*	3 hours
5	*Conduct Consolidation and Reorganization (Co/Plt)*	3 hours
ISSUE FRAGO, START STX C-II-2		
6	*Perform Cordon and Search*	6 hours
7	*Take Action on Contact / AAR*	3 hours
8	*Conduct an Infiltration (Co/Plt)*	6 hours
9	*Perform Linkup*	3 hours
10	*Perform Passage of Lines*	2 hours
11	*Occupy Assembly Area*	3 hours
START STX C-II-3		
12	*Disarm Belligerents*	3 hours
13	*Handle Captured Insurgents or Belligerents*	4 hours
14	*Apprehend/Detain Noncombatants*	4 hours
15	*Conduct Final AAR*	2 hours
	***Total Time**	**55 hours**

The time required to train an event varies based on the factors of METT-T and on the unit's training proficiency. Regardless, training is conducted to standards, not to time limitations. The following tasks are integrated and evaluated throughout the exercise:

- *Prepare for combat*
- *Maintain operations security*
- *Conduct tactical movement*
- *Perform consolidation*
- *Perform reorganization*
- *Perform stability and support operations*

* Time for intermediate AARs is excluded from this total; additional time is required if much of the exercise is conducted at night or during other limited visibility.

(4) The battalion commander issues an OPORD to the company commanders (Figure 5-9).

(Classification)

Copy no ___ of ___ copies
___ Battalion, ___ Infantry
Unit location or place of issue
Date-time group (with time zone)
Message reference number

OPERATION ORDER NO. _____
References: Map, series no. ___, sheet no. ___, edition ___
Time Zone Used Throughout the Order: ____

TASK ORGANIZATION:
___Company ___Company ___Company Battalion Control
Scout Platoon
81-mm Mortar Platoon
Antitank Platoon
1/C/58th ENGR (DS)
Battalion Trains

1. SITUATION.

a. **Enemy Forces**. Annex A (See intelligence overlay [prepared by brigade S-2].)

(1) *Disposition, Composition, and Strength.* The battalion is opposed by one company from the Dona Anna Militia Guards Battalion. This enemy company consists of 3 platoons with 35 to 40 Soldiers, equipped with M16s, AK-47s, and LMGs. The company headquarters is currently located in the town hall of CACTF. The company is at 90 percent strength and has indirect fire support assets available.

(2) *Capabilities.* The enemy possesses nonpersistent nerve agents and indirect fire assets capable of delivery. He also has some light armored vehicles in the area, but their location is undetermined. No enemy air threat exists. However, the enemy may be able to reinforce CACTF with an additional company, within two hours.

(3) *Most Probable Course of Action.* The enemy defends CACTF and continues using it as a base of operations and to consolidate weapons and material. His intent is to prevent the capture of CACTF and its support base, provided by its facilities and population.

(Classification)

Figure 5-9. Example battalion OPORD for FTX Attack.

(Classification)

b. **Friendly Forces.**

(1) (___) Brigade, (___) Division attacks at (date-time group) to seize the town of CACTF at (grid), to keep the enemy from using it as a base of operations. The intent is to secure the town and the road network in the area, to prevent the enemy from using it as a base of operations.

(2) (___) Battalion, (___) Infantry attacks to seize the road network at OBJ (____), vicinity (grid), at (date-time group) to prevent enemy forces from counterattacking along Route (____).

(3) (___) Battalion, (___) Infantry attacks at (date-time group), to seize the high ground at OBJ (____), vicinity (grid), and to establish defensive positions to block enemy routes in and out of CACTF.

2. **MISSION.** (___) Battalion, (___) Infantry attacks at (date-time group) to seize CACTF, vicinity (grid), NLT (date-time group), to prevent the enemy from using the town as an operations and logistics base.

3. **EXECUTION.**

a. **Concept of the Operation.** The battalion will accomplish its mission, which is to deny the enemy use of CACTF and its surrounding terrain and population, by seizing the town and controlling the key areas in the vicinity. We will conduct a feint in the west and then attack the decisive point within CACTF (OBJ [____]).

(1) *Maneuver.* (___) Company conducts a feint to deceive the enemy, then the main effort (*evaluated*) company seizes OBJ (____), to prevent the enemy from using CACTF as an operational and logistical base. (___) Company supports the main effort by seizing OBJ (____), in order to prevent the enemy from counterattacking the main effort. (___) Company seizes OBJ (____) to support the main effort's attack.

(2) *Fire Support.* Annex C. (See fire support.) Purpose of fires is to suppress the enemy and to obscure movement to CACTF. Priority of artillery fires is to (feinting) Company; on order, priority will shift to (supporting effort) Company. Priority of 81-mm mortar fires is to (*evaluated*) Company.

(3) *Engineering.* The purpose of the engineers is to breach obstacles during the attack, to help (*evaluated*) Company seize OBJ (____). Priority of support is to (*evaluated*) Company, (___) Company, and (___) Company in order. The battalion commander must authorize any booby traps, and point and hasty minefields.

(Classification)

Figure 5-9. Example battalion OPORD for FTX Attack (continued).

(Classification)

b. **Tasks to Maneuver Units.**

(1) *(Evaluated) Company.*

(a) Infiltrate using infiltration lane TANGO.

(b) Seize OBJ (____), vicinity (grid), NLT (date-time group).

(c) Be prepared to perform cordon and search of CACTF, after consolidation and reorganization.

(2) *(Supporting) Company.*

(a) Follow (main attack) Company and infiltrate on infiltration lane TANGO.

(b) Seize OBJ (____), vicinity (grid), NLT (date-time group).

(c) Coordinate the attack with *(evaluated)* Company, so that both units attack at the same time.

(d) Be prepared to perform cordon and search in the objective area after consolidation and reorganization.

(e) Be prepared to assume the mission of *(evaluated)* Company.

(3) *(Supporting) Company.*

(a) Infiltrate using infiltration lane ZEBRA.

(b) Conduct feint to OBJ (____), vicinity (grid), NLT (date-time group).

(c) Be prepared to establish defensive positions vicinity (grid) to prevent enemy movement to and from CACTF.

c. **Tasks to Subordinate Units.**

(1) *(Scout) Platoon.*

(a) During movement, determine location, strength, and disposition of enemy forces in CACTF.

(b) On order, screen the battalion flank from (grid) to (grid) to warn of the approach of enemy units from the southwest

(2) *Mortar Platoon.* Priority of mortar fire is to *(evaluated)* Company. Mortar platoon moves with *(evaluated)* Company, establishes a firing position vicinity (grid), and displaces on order.

(3) *Antitank Platoon.*

(a) Initially, over watch from (grid) to (grid) to secure the unit infiltration.

(b) On order, move vicinity (grid) to support *(evaluated)* Company's consolidation and reorganization.

(Classification)

Figure 5-9. Example battalion OPORD for FTX Attack (continued).

(Classification)

(4) *1/C/58 Engineers.*

(a) Move with *(evaluated)* Company to provide mobility support.

(b) Be prepared to reduce obstacles in CACTF during cordon and search mission.

d. **Coordinating Instructions.**

(1) Report when ready to execute an assault.

(2) Report any movement of enemy forces from the southwest.

(3) No rubbling is authorized, so minimize collateral damage to property.

(4) MOPP0 is in effect.

(5) Terrorist threat warning is red.

(6) Commanders brief back OPORD to battalion commander at battalion tactical CP at (date-time group).

(7) Rules of Engagement (ROE).

(a) Commanders take all steps necessary and appropriate for their units' protection.

(b) The minimum necessary force is used to control the situation.

(c) Leaders must take measures to minimize risk to civilians, without endangering the unit.

(d) Fire is returned directly to its source, not sprayed into a general area.

(e) Firing ceases when the threat is over.

(f) Anyone trying to surrender is allowed to do so.

(g) Civilians and property are treated with respect.

(h) WP can be used vicinity CACTF to aid in isolating the objectives. The requests for indirect fire, within the town, must be authorized by the battalion commander.

4. **SERVICE SUPPORT.** (See Annex D.)

5. **COMMAND AND SIGNAL.**

a. **Command.**

(1) The tactical CP moves with *(evaluated)* Company.

(2) The main CP is vicinity (grid) and displaces on order.

(3) The second in command remains in the battalion main CP.

(4) The rear CP is the alternate battalion main CP.

(Classification)

Figure 5-9. Example battalion OPORD for FTX Attack (continued).

(Classification)

b. **Signal**.

(1) Current SOI is in effect.

(2) Radio-listening silence is in effect for all elements except scouts, until prepared to attack.

ACKNOWLEDGE:
JONES
LTC
OFFICIAL:
SMITH
S-3

Annexes: A–Intelligence
B–Operations Overlay
C–Fire Support
D–Service Support

(Classification)

Figure 5-9. Example battalion OPORD for FTX Attack (continued).

(5) The battalion has destroyed the enemy operating in CACTF, and has seized the town. Two companies are defending against a limited counterattack. One company is in a blocking position. The (*evaluated*) company receives a FRAGO (Figure 5-10) to conduct a cordon and search operation. The purpose of the operation is to locate enemy personnel and equipment remaining in the CACTF.

Enemy personnel in squad-sized elements are reported by the civilian population to be hiding in the CACTF. Reports also indicate numerous weapons caches are being hidden by civilians collaborating with the enemy. Battalion has completed its mission and is preparing to conduct a cordon and search.

After completing this, the battalion exfiltrates to perform a passage of friendly lines. (Evaluated) Company moves to perform cordon and search vicinity OBJ (____), beginning (date-time group) to (date-time group), to locate and seize enemy personnel or equipment.

The (*evaluated*) company notifies the commander of the results of the search, and then notifies the commander when they (*evaluated* company) are ready to start movement. The company exfiltrates on order along Route Gold to perform linkup at (grid) and passage of lines at (grid). (See overlay for lanes.)

Figure 5-10. Example FRAGO for FTX Attack.

4. **Support Requirements**.

a. *Minimum Trainers/Evaluators*:. The company commander acts as the trainer-evaluator for the platoons. The battalion may direct the FTX internal evaluation along with a company evaluator and an FO evaluator. The platoon leaders and platoon sergeants are the primary trainers during the initial phases of the training. The company commander, the XO, the platoon leaders, and the platoon sergeants all act as trainers during the company phases of training. If the exercise is conducted by the battalion, either the battalion commander, XO, or S-3 acts as the trainer and evaluator, but additional battalion assets may be required to support the evaluation phase. The controllers or evaluators are positioned at critical locations throughout the exercise to observe the action. They must not interfere with the conduct of the operation, except to halt it when a safety violation occurs or to conduct an AAR.

b. *Vehicles/Communications*: Those organic to the company are used in the exercise. This includes one designated medical evacuation vehicle per platoon and at least one aidman on site.

c. *Opposing Force*:. The OPFOR ground force should be at least platoon-size. FTXs conducted by the company require internal resourcing; units should rotate to ensure total training. OPFOR tasks and standards are provided in each T&EO.

d. *Maneuver Area*: A training area 5 by 10 km which includes the CACTF is best for this exercise.

5. Evaluation.

Table 5-8 lists the T&EOs used to evaluate this STX.

Table 5-8. T&EOs used to evaluate STX Attack.

Task/Function	*Task Number*
Perform Cordon and Search	7-3-1105
Take Action on Contact	7-4-1107
Perform Passage of Lines	7-4-1125
Perform Linkup	7-4-1128
Conduct Tactical Movement	7-2-1342
Occupy Assembly Area	7-4-1136
Maintain Operations Security	7-4-1409
Prepare for Combat	7-4-1606
Conduct Consolidation and Reorganization (Co/Plt)	7-2-5027
Perform Cordon and Search	7-2-1105
Execute an Assault (MOUT)	7-2-1109
Perform Exfiltration	7-1-1146
Perform Infiltration	7-1-1147
Perform Reorganization	7-1-1608
Perform Consolidation	7-1-1609

MECHANIZED INFANTRY COMPANY OR TEAM FTX DEFEND

1. **Objective**. This FTX is designed to train the mechanized Infantry company or team in conducting defensive operations in urban terrain. It also provides platoons an opportunity to prepare for full-scale FTXs conducted and evaluated by the parent battalion. This exercise provides practice for the company commander, platoon leaders, squad leaders, and company headquarters personnel in planning, coordinating, and controlling combat operations.

2. **Interface**. This FTX is supported by the STX C-II-4, C-II-5, and C-II-6 tasks shown in Table 5-9. It is also supported by two warrior drills, *React to Contact* and *Break Contact*.

Table 5-9. Supporting tasks for company FTX.

STX C-II-4	*STX C-II-5*	*STX C-II-6*
Occupy Assembly Area	Perform Ambush	Operate a Checkpoint
Perform Tactical	Establish a Roadblock/Checkpoint	Interdict Smuggling Operations
Road March	Perform CBRN Operations	Defend Convoy
Perform Relief in Place	Conduct Tactical Movement	
Defend Built-Up Area/ Building	Perform Passage of Lines	
	Occupy Assembly Area	

3. **Training**.

a. *Guidance*. FM 7-0 and FM 7-1 provide training guidance, tips, and enhancers for planning and executing STXs. The trainer should review the individual, leader, and collective tasks to be performed during the STX to determine which require initial or refresher training.

b. *General Situation*.

(1) The battalion is ordered to conduct defensive operations in support of the brigade mission; the company is ordered to perform a relief operation and to prepare a defense in an urban area. The company remains in position for at least 36 hours before receiving a follow-on mission. Most civilians have been evacuated from the area, but a few remain. A terrorist threat exists as well as the possibility of enemy battalion-level offensive actions.

(2) This exercise should be conducted under various environmental conditions. The company also conducts operations under threat of CBRN attack.

c. *Special Situation.*

(1) The planning process for the defense mission begins when the company receives a warning order. The exercise ends when the company completes the passage of lines and occupies an assembly area. Figure 5-11 shows a graphic scenario of task performance for the FTX Defend.

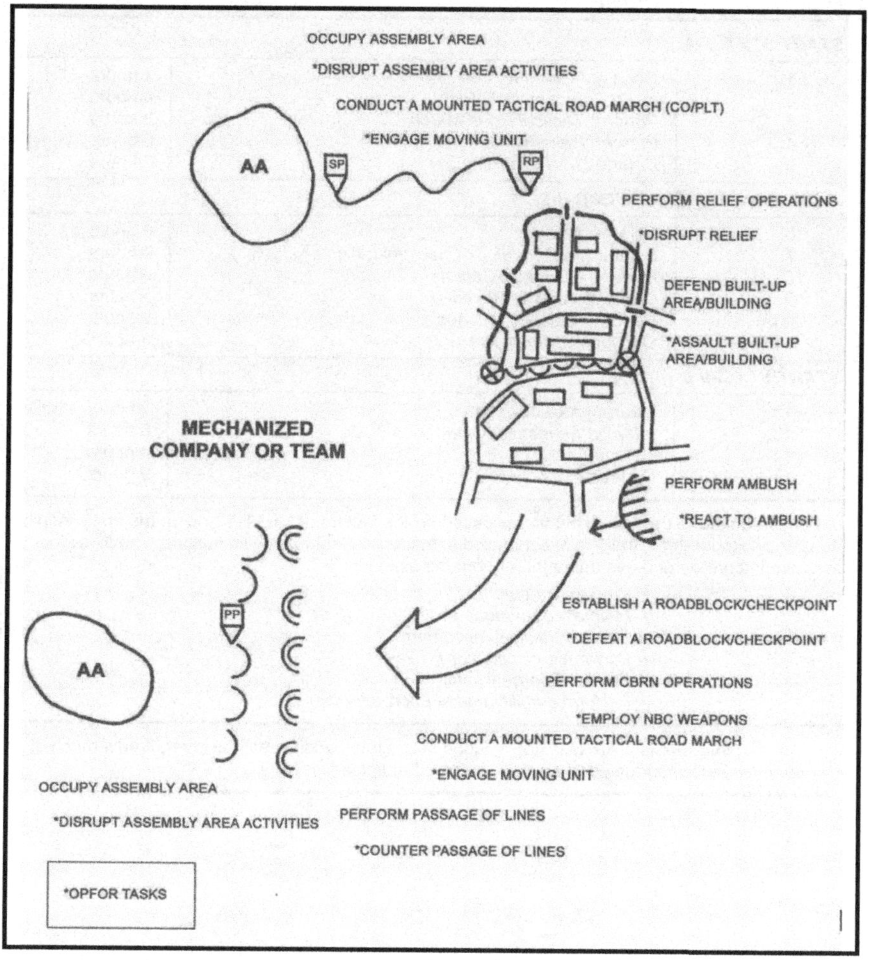

Figure 5-11. Graphic scenario of task performance for FTX Defend.

(2) Table 5-10 provides a recommended sequence for performance of T&EOs from DTMS and the time allotted for each portion of the STX, including AARs.

Table 5-10. Sequence of events and times for FTX Defend.

Event	*Action*	*Estimated Time*
START STX C-II-4		
1.	*Occupy Assembly Area*	2 hours
2.	*Issue Company OPORD*	4 hours
3.	*Perform Tactical Road March*	3 hours
4.	*Perform Relief Operations AAR*	6 hours
5.	*Defend MOUT/Building*	24 hours
ISSUE FRAGO, START STX C-II-5		
6.	*Perform Ambush*	6 hours
7.	*Establish a Roadblock/Checkpoint AAR*	6 hours
8.	*Perform CBRN Operations*	6 hours
9.	*Conduct Tactical Movement*	3 hours
10.	*Perform Passage of Lines*	2 hours
11.	*Occupy Assembly Area*	3 hours
START STX C-II-6		
12.	*Operate a Checkpoint*	2 hours
13.	*Interdict Smuggling Operations*	6 hours
14.	*Defend Convoy*	4 hours
15.	*Conduct Final AAR*	2 hours
The time required to train an event varies based on the factors of METT-T and on the unit's training proficiency. Regardless, training is conducted to standards, not to time limitations. The following tasks are integrated and evaluated throughout the exercise: ▪ *Prepare for combat* ▪ *Maintain operations security* ▪ *Conduct tactical movement* ▪ *Perform consolidation* ▪ *Perform reorganization* ▪ *Perform stability and support operations*		
* Time for intermediate AARs is not included in this total; additional time is required if much of the exercise is conducted at night or during other limited visibility.		

(3) The battalion commander issues the defense OPORD to the company (Figure 5-12).

(Classification)

Copy no ___ of ___ copies
___ Battalion, ___ Infantry
Unit location or place of issue
Date-time group (with time zone)
Message reference number

OPERATION ORDER NO. _____

Reference: Map, series no. ___, sheet no. ___, edition ___
Time Zone Used Throughout the Order: ____
TASK ORGANIZATION:
___Co ___Tm ___Tm (-) Battalion Control
1/__/__/AR 2/__/__/AR Scout Platoon
120-mm Mortar Platoon
Team Tank E Co(-) 1/C/1-222 ADA (Stinger) (DS)
1/__/__/Inf C/58 ENGR (DS)
2/__/__/Inf 1/1 (GSR) TM A/58 MI Bn
ATK Force B (1/_/_/Inf)
Battalion Trains

1. SITUATION.

a. **Enemy Forces**. Annex A (see intelligence overlay [prepared by brigade S-2]).

(1) *Composition, Disposition, Strength*. The company is opposed by the 1st Dragoons Regiment, 60th Guards Division. This regiment is at 85 percent strength and has received engineer and artillery reinforcements. The enemy is equipped with an assortment of former Soviet and Warsaw pact small arms, machine guns, and armored vehicles including T-62 tanks.

(2) *Capabilities*. The enemy has established defensive positions that follow a line from vicinity (grid) to vicinity (grid). Enemy armored reconnaissance units are operating in the more open terrain to the southwest of CACTF. The enemy has been observed moving forces in depth throughout the area in preparation for an offensive action.

(3) *Probable Course of Action*. The enemy's most probable course of action will be to begin offensive actions within the next 24 to 36 hours. Enemy encountered includes numerous reconnaissance elements trying to find the company's vulnerabilities and a main body trying to exploit any weakness.

(Classification)

Figure 5-12. Example battalion OPORD for mechanized company or team STX Defend.

(Classification)

b. **Friendly Forces.**

(1) (___) Brigade, (___) Division relieves (___) Brigade in place at (date-time group), to defend in sector from (grid) to (grid) to (grid) to (grid), in order to secure the corps counterattack axis of advance, (grid). The intent is to secure the road network around CACTF to prevent the enemy from impeding the movement of the (___) Brigade, (___) Division.

(2) TF (__) is the main effort in the defense and, NLT (date-time group), holds key terrain vicinity (grid) on our left, to prevent an enemy bypass of the brigade left flank.

(3) TF (__), on our right, defends along Route 360, vicinity (grid) to (grid) to (grid) to (grid), NLT (date-time group), to prevent an envelopment of the right flank. They conduct a supporting attack at (date-time group) to seize the high ground at OBJ OAK, vicinity (grid), to establish defensive positions that prevent the enemy armored reconnaissance units from interfering with the seizure of the pass at (grid) or the attack of the (___) Battalion, (___) Infantry, Mechanized.

2. **MISSION**. TF (____) defends NLT (date-time group), to retain the town of CACTF and its surrounding road network and allow brigade to prevent enemy interference with corps counterattack.

3. **EXECUTION.**

a. **Concept of the Operation**. Annex B (see operations overlay).

(1) Maneuver. Company (main effort) retains the main intersection (traffic loop) in CACTF, to prevent enemy movement through the sector. Team (*evaluated*) blocks movement west of 3d Street, to prevent the enemy from enveloping the main effort. The counterattack force is committed as needed. Team (__) blocks movement east of Oak Street, to canalize the enemy into EA (____). Team (__) defends from BP1, to prevent a coordinated attack against the main effort. E Company occupies BPs, to destroy enemy vehicles and protect the TF flanks.

(2) Fire Support. (See Annex C.) Fires are used to suppress the enemy and obscure our movement during the relief operation. Once positions are occupied, fires are used to suppress enemy overwatch positions, disrupt attempts to breach obstacles, and separate enemy Infantry from their supporting carriers. Priority of CAS and indirect fire is initially to Team (*evaluated*), during the relief operation, and then to Team (main effort). Teams (main effort) and (*evaluated*) each have two priority 155-mm targets. Each team may make one priority target an FPF.

(Classification)

Figure 5-12. Example battalion OPORD for mechanized company or team STX Defend (continued).

(Classification)

(3) Counter-Air Operations. Priority of protection is to the TF trains and main CP. Air defense weapons status is free; air defense warning status is yellow.

(4) Intelligence. Priority of collection is to identify and locate the C2 elements of the MRR and MRB, enemy tank battalions, and the enemy assault detachments' actual axes of advance.

(5) Engineering. The purpose of obstacles is to turn the enemy into restrictive terrain and engagement areas. Priority of support is to Team (main effort), Team (*evaluated*), Team (__), and Team (__), in that order.

b. **Tasks to Maneuver Units.**

(1) Team (Main Effort).

(a) Sector of defense includes the east side of Front Street, the south side of 1st Street, the west side of 3d Avenue area, and the north side of 3d Street.

(b) Team (main effort) relieves Team (__), TF (__).

(2) Team (Evaluated).

(a) Sector includes the east side of 3d Avenue, the south side of 1st Street, the west side of 6th Avenue, and the north side of 3d Street.

(b) Team (*evaluated*) relieves Team (__), TF (__).

(3) Team (__).

(a) Sector includes the east side of Oak Street, the south side of 1st Street, the east side of Front Street, and the north side of 3d Street.

(b) Team (__) relieves Team (__), TF (__).

(c) Team (__) coordinates directly with right flank team of TF (__), to ensure mutually supporting fires.

(4) Team (__)

(a) Occupy BP 1 vicinity (grid)

(b) Be prepared to occupy BP 2 (grid) to engage enemy in EA (_____).

(5) Company E.

(a) Occupy BP 2-1 (grid) and move on order to BP 2-2 (grid).

(b) Provide antiarmor fires to destroy enemy tanks in EA (_____).

(6) Counterattack B (Reserve).

(a) Occupy AA Black, vicinity (grid).

(b) Reconnoiter routes into TF (main effort) sector.

(Classification)

Figure 5-12. Example battalion OPORD for mechanized company or team STX Defend (continued).

(Classification)

(3) Counter-Air Operations. Priority of protection is to the TF trains and main CP. Air defense weapons status is free; air defense warning status is yellow.

(4) Intelligence. Priority of collection is to identify and locate the C2 elements of the MRR and MRB, enemy tank battalions, and the enemy assault detachments' actual axes of advance.

(5) Engineering. The purpose of obstacles is to turn the enemy into restrictive terrain and engagement areas. Priority of support is to Team (main effort), Team (*evaluated*), Team (__), and Team (__), in that order.

b. **Tasks to Maneuver Units.**

(1) Team (Main Effort).

(a) Sector of defense includes the east side of Front Street, the south side of 1st Street, the west side of 3d Avenue area, and the north side of 3d Street.

(b) Team (main effort) relieves Team (__), TF (__).

(2) Team (Evaluated).

(a) Sector includes the east side of 3d Avenue, the south side of 1st Street, the west side of 6th Avenue, and the north side of 3d Street.

(b) Team (*evaluated*) relieves Team (__), TF (__).

(3) Team (__).

(a) Sector includes the east side of Oak Street, the south side of 1st Street, the east side of Front Street, and the north side of 3d Street.

(b) Team (__) relieves Team (__), TF (__).

(c) Team (__) coordinates directly with right flank team of TF (__), to ensure mutually supporting fires.

(4) Team (__)

(a) Occupy BP 1 vicinity (grid)

(b) Be prepared to occupy BP 2 (grid) to engage enemy in EA (____).

(5) Company E.

(a) Occupy BP 2-1 (grid) and move on order to BP 2-2 (grid).

(b) Provide antiarmor fires to destroy enemy tanks in EA (____).

(6) Counterattack B (Reserve).

(a) Occupy AA Black, vicinity (grid).

(b) Reconnoiter routes into TF (main effort) sector.

(Classification)

Figure 5-12. Example battalion OPORD for mechanized company or team STX Defend (continued).

(Classification)

(c) Priority for planning:
- Battle position to reinforce main effort.
- Battle position to counterattack into Team (main effort) sector.
- Battle position to counterattack into Team (*evaluated*) sector.

(1) Scout Platoon.

(a) Determine location, strength, and disposition of enemy forces.

(b) Perform a screen of the TF's right flank from (grid) to (grid) beginning (date time group) to prevent enemy infiltration of CACTF.

c. **Tasks to Subordinate Units.**

(1) Mortar Platoon.

(a) Priority of fire to TF (main effort). Move with TF (main effort) and establish firing position vicinity (grid).

(b) Displace on order to (grid).

(2) 1/C/1 222 ADA. Coordinate locations with S-3 NLT (date time group).

(3) C/58 Engineers.

(a) Coordinate the engineer effort with the team commanders NLT (date time group).

(b) Provide an obstacle clearing team to Counterattack Force B, NLT (date time group).

(4) 1/1 (GSR) TM A/58 MI. Coordinate location of assets with the task force S-3 NLT (date time group).

d. **Coordinating Instructions.**

(1) Selective rubbling is authorized, but destruction of more than one building requires TF commander approval.

(2) Destruction of subterranean approaches requires TF commander approval.

(3) Commanders brief back OPORD to TF commander at the TF tactical CP, vicinity (grid), at (date time group).

(4) MOPP1 is in effect as of (date time group).

(5) Decontamination site established vicinity Miller Pond (grid).

(Classification)

Figure 5-12. Example battalion OPORD for mechanized company or team STX Defend (continued).

(Classification)

4. SERVICE SUPPORT. (See Annex D.)

a. **General.**
(1) Combat trains located vicinity (grid).
(2) Field trains located in the BSA, vicinity (grid).

b. **Materiel and Services.**
(1) Supply.
(a) Class I. Ration cycle C-A-C. Team establishes MRE cache sufficient for 48 hours in sector.
(b) Class III. Resupply available on request. Class III located in combat trains.
(c) Class IV. Materials are allocated as shown in Annex D.
(d) Class V. Each team establishes ammunition caches in sector, but should also have two basic loads of ammunition on hand.
(2) Transportation. The main streets are capable of two way heavy vehicle traffic (class 60). Secondary streets may be one way and of lighter construction (class 50).
(3) Maintenance.
(a) Priority by unit: Team (main effort), Team (*evaluated*), Team (__), Company E, Team (__).
(b) Priority by vehicle: M1A1, M2/M3, M113.

c. **Medical Evacuation and Hospitalization.**
(1) TF aid station located vicinity (grid).
(2) Civilian casualties should be transported by nonmilitary means to clinic, located vicinity (grid).

d. Personnel.
(1) No replacements for the next 96 hours.
(2) EPW collection point located vicinity (grid).
(3) Mortuary affairs point located in BSA, vicinity (grid).

e. **Civil-Military Cooperation.**
(1) Curfew in effect 1800 to 0800 daily.
(2) Majority of civilians have been evacuated. Those remaining have been notified of the risk involved.

(Classification)

Figure 5-12. Example battalion OPORD for mechanized company or team STX Defend (continued).

(Classification)

5. COMMAND AND SIGNAL.

a. **Command.**
(1) Brigade main CP located vicinity (grid).
(2) TF tactical CP located vicinity (grid).
(3) TF main CP located vicinity (grid).
(4) TF commander located at tactical CP.
(5) TF XO located at main CP.

b. **Signal.**
(1) SOI index 1 9 in effect.
(2) Primary means of communication is wire, FM, messenger in that order.
(3) Building and obstacle marking IAW TF SOP.

ACKNOWLEDGE:

WILLIAMS
LTC

OFFICIAL:

SMITH
S-3

Annexes: A–Intelligence
B–Operations Overlay
C–Fire Support
D–Service Support

(Classification)

Figure 5-12. Example battalion OPORD for mechanized company or team STX Defend (continued).

(3) The TF has retained the north half of the CACTF after the enemy attack. The enemy was defeated but has established squad-sized to platoon-sized strongpoints in the southern half of the CACTF. The enemy is expected to receive reinforcements within the next 12 to 24 hours. Theprevent enemy reinforcements from moving to the strongpoints to strengthen and resupply them, TF headquarters orders the team to conduct platoon-sized ambushes. They also order the team to prepare to establish a roadblock, to control refugee flow through the area of operations, before the team moves out of the area. Headquarters issues a FRAGO (Figure 5-13).

```
Numerous enemy strongpoints are located to your front, vicinity (grid), (grid), and (grid). We expect these strongpoints to receive reinforcements within the next 12 to 24 hours.

Your mission is to establish platoon ambushes on the likely avenues of approach to the strongpoints, NLT (date-time group), in order to prevent reinforcements or resupply to these strongpoints.

Notify the TF commander when you are prepared to move.

Coordinate with TF scouts for guides. On order, after return from ambush missions, establish roadblocks at (grid) and (grid). Be prepared to move within the next 36 to 48 hours to conduct further follow-on missions.
```

Figure 5-13. Example FRAGO for FTX Defend.

4. **Support Requirements.**

a. *Minimum Trainers/Evaluators*: This exercise should be conducted for practice with the company commander as the trainer-evaluator for his platoons. The battalion may direct the FTX (internal evaluation) using a company evaluator and an FO evaluator. The platoon leaders and platoon sergeants are the primary trainers during the initial phases of the training. The company commander, the XO, the platoon leaders, and the platoon sergeants act as trainers during the company phases of training. If the exercise is conducted by the battalion, either the battalion commander, the XO, or the S-3 acts as the trainer and evaluator, but may require additional assets from the battalion to support the evaluation phase. The controllers or evaluators must be positioned at critical locations throughout the exercise, to observe the actions. They must not interfere with the conduct of the operation, except to halt it when a safety violation occurs or to conduct an AAR.

b. *Vehicles/Communications*: Those organic to the company are used.

c. *Opposing Force*:. The OPFOR ground force should be at least platoon-size. FTXs conducted by the company require internal resourcing, and units should be rotated to ensure total training. T&EOs provide training tasks and standards for the OPFOR.

d. *Maneuver Area*: A training area at least 5 by 15 km is best for this exercise.

e. *Consolidated Support Requirements*: This exercise requires the items shown in Table 5-11.

f. *Barrier materials.* Sufficient Class IV must be made available to fortify positions, IAW FM 3-06.11.

Table 5-11. Ammunition and pyrotechnics support requirements for a six-day FTX.

Ammunition	*DODAC*	*Per Plt/Co/Bn FTX*
5.56-mm blank	1305A080	120 rounds per rifle
5.56-mm SESAMS		120 rounds per rifle
5.56-mm blank	1305A075	200 rounds per M249 LMG
5.56-mm SESAMS		200 rounds per M249 LMG
7.62mm	1305A111	200 rounds per BFV/M1
		400 rounds per M60/240B
ATWESS cartridge	1370L367	6 (Viper) (for LAW) 18 per Dragon 14 per TOW system 9 per OH-58D/AH-64
Pyrotechnic device	1370L602	6 per M1 2 per target
Hand grenade fuze (M228)	1330G878	150 per battalion
Stun grenade M84	1330GG09	150 per battalion
Smoke grenade, HC		150 per battalion
Smoke pot		6 per battalion

5. **Evaluation.**

Table 5-12 lists the DTMS T&EOs used to evaluate this STX.

Table 5-12. T&EOs used to evaluate STX Attack.

Task/Function	Task Number
Defend MOUT/Building	7-3/4-1118
Conduct a Dismounted Tactical Road March (Co/Plt)	7-2-1189
Conduct a Mounted Tactical Road March (Co/Plt)	7-2-1198
Perform Relief Operations	7-3/4-1124
Perform Passage of Lines	7-3/4-1125
Conduct Tactical Movement	7-2-1342
Occupy Assembly Area	7-3/4-1136
Establish a Roadblock/Checkpoint	7-3/4-1401
Perform CBRN Operations	7-3/4-1406
Maintain Operations Security	7-3/4-1409
Prepare for Combat	7-3/4-1606
Conduct Consolidation and Reorganization	7-2-5027
Perform Ambush	7-2-1106

TARGETRY

5-10. Table 5-13 lists the combination of precision and nonprecision targets to support the diversity of scenarios required to support combined arms and branch specific lane training. Precision targets, less than 50 meters, are used in and around buildings within the CACTF and reinforce precision marksmanship techniques described in FM 3-22.9 (Table 5-13).

Table 5-13. CACTF target requirements.

Event	Number Of Lanes	Human Targets		Vehicle Targets	Totals	
					HT	VT
Company/Platoon STX Offense	3/9	Precision	Nonprecision	9	45	9
		15	30			

CONSIDERATIONS FOR TRAINING WITH ARMOR

5-11. Built-up areas consist mainly of man-made features such as buildings, streets, and subterranean systems. These features of urban terrain create a variety of tactical problems and possibilities. In order to ensure that the tank platoon can operate effectively in the urban environment, the platoon observation and direct-fire plans must address the ground-level or surface fight (in streets and on the ground floor of buildings), the aboveground fight (intrasurface and supersurface), and the subterranean fight (Figure 5-14). Infantry can assist the tank platoon by—

- Locating targets for tanks to engage.
- Destroying antitank weapons.
- Assaulting enemy positions and clearing buildings with tank support.
- Protecting tanks from antitank fires.

Figure 5-14. Urban battlespace.

TRANSPORTING INFANTRY

5-12. At times, the tank platoon must transport Infantrymen (Figure 5-15). They only do so when not expecting contact. If the platoon is moving as part of a larger force, and if it is tasked to provide security for the move, then the lead section or element does not carry Infantry.

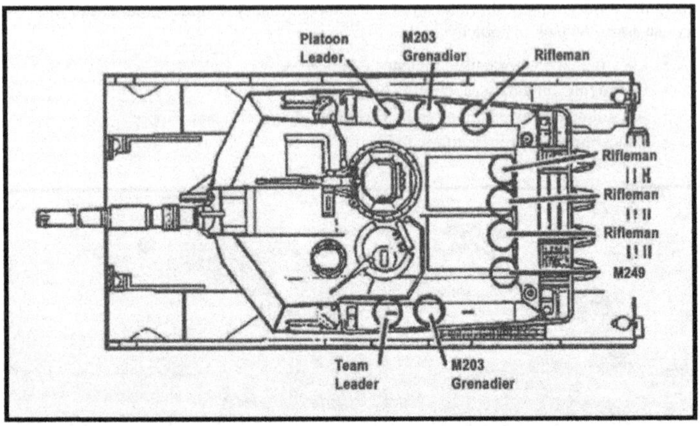

Figure 5-15. Example positions for Infantry riding on a tank.

5-13. Infantry and Armor leaders must observe the procedures, precautions, and considerations in Figure 5-16 when Infantrymen ride on tanks:

- Infantry should thoroughly practice mounting and dismounting procedures and actions on contact.
- Infantry must always alert the TC before mounting or dismounting. They must follow the commands of the TC.
- Infantry platoons should be broken down into squad-sized groups, similar to air assault chalks, with the Infantry platoon leader on the Armor platoon leader's vehicle and the Infantry PSG on the Armor PSG's vehicle.
- Platoon leaders, PSGs, and team leaders should position themselves near the TC's hatch, using the external phone (if available) to talk to the TC and relay signals to the unit.
- If possible, the lead vehicle should not carry Infantrymen. Riders restrict turret movement and are more likely to be injured or killed on initial contact.
- Whenever possible, Infantrymen should mount and dismount over the left front slope of the vehicle. This procedure ensures that the driver can see the Infantrymen and that the Infantrymen DO NOT pass in front of the coax machine gun. Infantrymen must ensure that they remain behind the vehicle's smoke grenade launchers. This will automatically keep them clear of all weapon systems.
- Infantrymen must always have three points of contact with the vehicle; they must watch for low-hanging objects such as tree branches.
- All Infantrymen should wear hearing protection.
- Infantrymen should not ride with anything more than their battle gear. Rucksacks and B-bags should be transported by other means.
- Infantrymen should scan in all directions. They may be able to spot a target the vehicle crew does not see.

Figure 5-16. Precautions for Infantrymen riding on a tank.

5-14. Passengers should be prepared to take the actions on contact shown in Figures 5-17 and Figure 5-18.

- Wait for the vehicle to stop.
- At the TC's command, dismount immediately (one fire team on each side). ***Do not*** move forward of the turret.
- Move at least 5 meters to the sides of the vehicle.
  Do not move behind or forward of the vehicle.
- ***Do not*** move in front of vehicles unless ordered to do so. Main gun discharge overpressure can inflict severe injury or death to forward dismounts.
- ***Do not*** dismount a vehicle unless ordered or given permission to do so.
- ***Do not*** dangle arms or legs, equipment, or anything else off the side of a vehicle; they could get caught in the tracks, causing death, injury, or damage to the equipment or vehicle.
- ***Do not*** place too many riders on the vehicle.
- ***Do not*** fall asleep when riding. The warm engine may induce drowsiness; a fall could be fatal.
- ***Do not*** smoke when mounted on a vehicle.
- ***Do not*** stand near a moving or turning vehicle at any time. Tanks have a deceptively short turning radius.

Figure 5-17. Actions on contact for Infantrymen riding on a tank.

DANGER

1. The overpressure from the tank 120-mm cannon can kill a dismounted Infantryman within a 90-degree arc extending from the muzzle of the gun tube out to 200 meters.
2. From 200 to 1,000 meters along the line of fire, on a frontage of about 400 meters dismounted Infantry must be aware of the danger from discarding sabot petals, which can kill or seriously injure Soldiers.

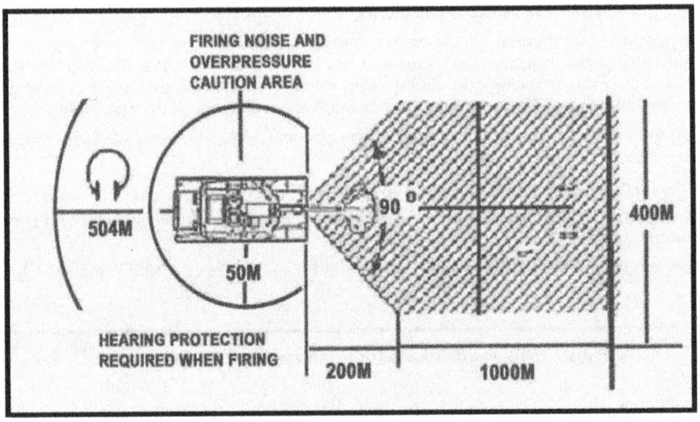

Figure 5-18. Danger areas around a tank when firing the 120-mm main gun.

ADDITIONAL CONSIDERATIONS

5-15. Additional considerations and preparations for transporting Infantrymen include the following:

- Armor--
  - Uses main gun fire to reduce obstacles or entrenched positions for the Infantry.
  - Takes directions from the Infantry ground commander (platoon leader/PSG/squad leader) to support their fire and maneuver.
  - Provides reconnaissance by fire for the Infantry.
  - -- Should know and understand how the Infantry clears buildings, how they mark cleared buildings, casualty evacuation plan, signal methods, engagement criteria for tank main gun, front line trace reporting, ground communication from the tank with the dismounted Infantry.
  - -- Uses its night vision capability to augment and supplement the Infantry's night vision capability.
- Infantry--.
  - -- Provides real time information for the tank crewmen to help them overcome tank noise and the lack of ground situational understanding.
  - -- Provides reconnaissance and fire direction of enemy positions for main gun attack.

- Considerations for dismounted tank security include--
  - -- Tank crewman should rehearse the mounting and dismounting of Infantry from their vehicle, briefing the Infantrymen on safety procedures for the vehicle and weapon systems.
  - -- Tank commanders need to rehearse communicating with dismounted Soldiers by way of TA-1 and DR-8 in the bustle rack.
- Vehicle preparation for combat in urban terrain should cover--
  - -- Keep at least one ballistic shield to the "Dog House" closed. Most engagements will be under boresight range, and the battlesight technique will suffice.
  - -- Place sandbags around antenna connections and electrical wiring on the turret top.
  - -- Place extra coax ammunition inside the turret.
  - -- Remove any highly flammable products from the outside of the vehicle and from the sponson boxes.

VEHICLES, WEAPONS, AND MUNITIONS

5-16. Numerous factors related to vehicles and their organic weapons and munitions affect the tank platoon's urban operation planning and execution, including the following:

- The preferred main gun rounds in the urban environment are HEAT, MPAT (ground mode), and MPAT-OR. These all perform much better than sabot rounds against bunkers and buildings.
- HEAT ammunition will open a larger hole in reinforced concrete or masonry structures than MPAT or MPAT-OR. Both MPAT and MPAT-OR, however, offer greater incapacitation capability inside the structure.
- HEAT ammunition arms approximately 60 feet from the gun muzzle. It loses most of its effectiveness against urban targets at ranges of less than 60 feet.
- MPAT and MPAT-OR rounds arm approximately 100 feet from the muzzle of the gun. Because of the shape and metal components of the projectiles, however, this ammunition remains effective at ranges of less than 100 feet.
- Sabot petals, including those on MPAT and MPAT-OR, endanger accompanying Infantry elements. They create a hazard area extending 70 meters on either side of the gun-target line out to a range of 1 kilometer.
- The tank's main gun can depress only to -10 degrees and can elevate only to +20 degrees. This creates considerable dead space for the crew at the close ranges that are typical in the urban environment.
- The external M2 HB machine gun can deliver a heavy volume of suppressive fire and penetrate light construction, buildings and most barricades. The M2 HB MG can elevate to +36 degrees; however, the TC must be unbuttoned to fire the M2 on the M1A2 or M1A2 SEP.
- The M240 coax machine gun can effectively deliver suppressive fires against enemy personnel and against enemy positions that are behind light cover.
- The loader's M240 machine gun can effectively deliver suppressive fire against enemy personnel and against enemy positions that are behind light cover; however, the loader must be unbuttoned to operate it. This weapon may be dismounted and used in a ground role if units are equipped with the M240 dismount kit.

- When buttoned up, the tank crew has limited visibility to the sides and rear and no visibility to the top. Figures 5-19 and 5-20 show the dead spaces associated with tank operations in an urban environment.
- FM 3-20.12 explains special uses for tank-mounted machine guns in the urban environment.

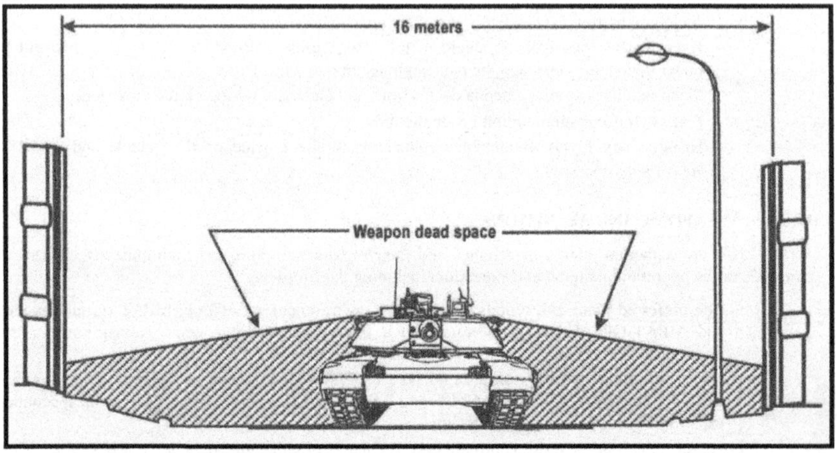

Figure 5-19. Tank weapon dead space at street level.

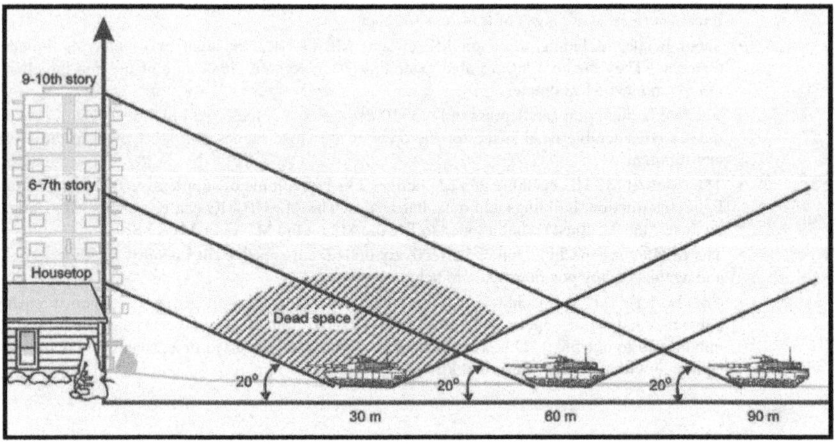

Figure 5-20. Tank main gun and coax dead space above street level.

Tank Platoon Command and Control in Urban Operations

5-17. Communications problems, fire control, proximity and visibility, personnel factors, ROE, ROI, civilians, and slow-pace urban operations all affect the tank platoon's urban operations planning and execution:

Communications Problems

5-18. The low-level task organization that may take place during urban operations requires elements to establish additional communications links, which can be disrupted by buildings and other urban terrain features.

Fire Control

5-19. Extensive direct fire planning and restrictive fire control measures are an absolute requirement in urban operations.

Proximity and Visibility

5-20. Friendly elements often must operate in confined and restrictive areas during urban operations, and they may not be able to see other nearby friendly forces. These factors significantly increase the danger of fratricide.

Personnel Factors

5-21. Urban operations impose significant, and often extreme, physical and psychological demands on Soldiers and leaders.

ROE, ROI, and Civilian Considerations

5-22. The ROE or ROI may restrict the use of certain weapon systems and TTP. As an integral part of urban operations, noncombatants create special operational problems. Thedeal with these concerns, units operating in urban terrain must know how to effectively employ linguists and counterintelligence and civil affairs teams.

Slow Pace Urban Operations

5-23. This operation usually prevents the tank platoon from taking full advantage of the speed and mobility of its tanks.

Appendix A

Training Tips

The effectiveness of urban operations training depends on the imagination and thoroughness applied to a training plan. All trainers, from training managers to junior leaders, should prepare their units to fulfill their urban operations mission. They can employ procedures that are based on training doctrine and proven techniques. This appendix discusses several techniques for conducting urban operations training. Trainers make the choice based on the training needs and available resources.

CONCURRENT TRAINING

A-1. Concurrent training allows groups of Soldiers to train on different tasks at the same time. This training requires a plan for rotating Soldiers between the stations that teach different tasks. This procedure is referred to as round-robin training. Concurrent training is an efficient training technique since it makes the best use of training time and facilities. Concurrent training in the urban operations training facilities works best with performance-oriented training. It also requires detailed planning, control, coordination, and assistant trainers. Each station should conduct training that is relevant to, and supports the training the unit is conducting.

URBAN ASSAULT COURSE AND BREACH FACILITY

A-2. These facilities are well suited for conducting concurrent training. After mastering the tasks at one station, a group of Soldiers rotates to the next station in the sequence. Theavoid bottlenecks at any of the stations, trainers plan instructional blocks that last about the same length of time. Concurrent training is often conducted with dedicated instructors at each station. The *advantage* of this technique is that it ensures uniformity of performance standards and saves on trainer preparation time. The *disadvantage* is that the Soldier is not always trained by his leader. Therefore, when feasible, small-unit leaders should rotate with their Soldiers from station-to-station so they can take part in the training. This technique demands prior planning, preparation, and instructors that can teach.

URBAN ASSAULT COURSE AND COMBINED ARMS COLLECTIVE TRAINING FACILITY

A-3. These facilities are well suited for conducting concurrent exercises. Concurrent exercises are variants of concurrent training. In a concurrent exercise, elements of the same unit receive different missions. These missions correspond to each element's training objective for that exercise. This technique can be used during FTXs, STXs, or CFXs. For example, one unit attacks; a second unit supports by fire; a third unit acts as a reserve. Also, elements of the same unit can challenge each other. For example, if a trainer has been training his unit in urban offense and defense operations, he can create a scenario in which one force is required to attack the other. This technique allows the unit to train with an OPFOR without an extra investment of resources. Leaders must remember to carefully plan the force ratios between the OPFOR and counterattacking forces.

DEMONSTRATIONS

A-4. Trainers must demonstrate the correct way to perform the tasks that the Soldiers will be required to perform. A demonstration is more effective when the trainer uses hand-on techniques. A live demonstration is an effective teaching method since it holds Soldiers' interest and they are able to learn with their eyes. The UAC and the Breach Facility are ideal facilities for teaching individual and collective tasks using demonstrations. The trainer must ensure that each Soldier. Performs the task being taught to guarantee that the soldier has learned. Key points should be explained to the satisfaction of each Soldier before progressing to the next step. The trainers can use video cameras to film the action occurring inside the buildings to show the Soldiers the correct way to perform a task or use the same action to explain a better technique to accomplish the same task. This technique is very effective because the Soldiers can watch the demonstrations on a remote monitor, and trainers can replay the videotape as often as required to reinforce his training objective.

EXERCISES

A-5. Training exercises are used to develop tactical SOPs, teamwork among the participants, and Soldier confidence in his leaders and equipment. Table A-1 matches training exercise types to appropriate training facilities.

Table A-1. Training exercise matrix.

EXERCISE	*CACTF*	*SHOOT HOUSE*	*SIMULATIONS*	*LOCAL TOWN*
MAPEX	X		X	X
TEWT	X			X
CPX	X		X	X
CFX	X			X
BSX			X	
LFX		X		
STX	X	X		X
FTX	X			X

MAP EXERCISE

A-6. A MAPEX allows participants to plan and execute urban operations using maps, overlays, sand tables, photographs, or models. MAPEXs develop teamwork and skills in terrain analysis, planning, coordination, order writing, and employment of urban operation tactics and are more suited for battle staff and leaders.

- Using a model of the CACTF is an excellent training aid for instructing Soldiers and leaders. Gather leaders around the map or model and have them devise a tactical plan. Use figures or markers to represent friendly units and the OPFOR. Create tactical situations that challenge participants to react as they would in real combat.
- Incorporate direct fire, casualties, CBRN, logistics, and civilians into the exercise. Ensure the ROE for each exercise is clear to participants.
- Have plenty of detailed maps available for use in the MAPEX and in the planning and execution of subsequent CFX operations.

Tactical Exercise Without Troops

A-7. A TEWT focuses on leader and staff tasks. The senior trainer selects the terrain for the urban TEWT. Either the CACTF or a real village, town, or city is appropriate. The procedure for conducting a TEWT is as follows:

- Create a scenario and enemy situation, and assign missions.
- Allow the leaders to devise their reconnaissance plans and implement them on the terrain.
- Require the leader to brief subordinates while walking the terrain.
- Ensure their plans include the following tasks:
  - -- Analyze the terrain.
  - -- Employ units according to the terrain analysis.
  - -- Emplace weapons.
  - -- Devise a unit plan.
- Conduct an AAR.

Command Post Exercise and Battle Simulation Exercise

A-8. CPXs and BSXs allow leaders and battle staffs to perform their tasks in a unit CP, where they receive information as reports requiring certain reactions. This type of training can be achieved through the constructive simulations training portion of the urban operations training strategy using JANUS or JCATS. A company CP is usually associated with a battalion level or higher CPX, but a company can conduct its own CPX. The headquarters section can set up a CP in a CACTF building, displace to another building, and repeat the exercise until they meet the required standards. The section can practice hiding wire on pavement, stringing it between buildings, and running it through sewers. The CP can practice receiving reports, posting situation maps, and sending reports forward to higher headquarters. The CPX/BSXs are most effective in achieving the following general training objectives:

-- Preparing plans.
-- Issuing orders.
-- Building teams.
-- Reconnoitering, selecting, and tactically occupying CP locations.
-- Establishing and using communications.
-- Preparing and sending reports.
-- Coordinating between subordinate elements.

Command Field Exercise

A-9. The CFX is used to train commanders, staffs, and junior leaders. The CACTF or a local town can accommodate this type of exercise. However, the unit commander must go through the correct procedures when conducting training off federal property. The CFX employs all of a unit's command and control assets, with a number of subordinate elements reduced according to a set scale. For example, one Soldier might represent an entire squad, one vehicle a platoon, or one gun an entire battery.

Preliminary Training

A-10. Preliminary training, such as CPXs, TEWTs, MAPEXs, and BSXs, should precede the CFX. Units conduct operations, and controllers resolve battle outcomes. The exercise ends with an AAR. Fewer controllers are needed in a CFX than in an FTX. However, if the number of unit functions increases, the number of controllers must increase also.

Planning Procedures

A-11. Combat in urban areas entails centralized planning with decentralized execution.

Centralized

A-12. Centralized planning can be practiced in a CFX; however, it can be practiced more efficiently in a CPX, TEWT, MAPEX, or BSX.

Decentralized

A-13. Decentralized execution means that platoon, squad, and team leaders lead the battle in an urban area. However, because element participation is scaled down in a CFX, trainers usually find STXs, FTXs, and TEWTs more useful.

LIVE-FIRE EXERCISE

A-14. Chapter 3 and local range regulations provide specific guidance for conducting LFXs at the shoot house. Live-fire training allows Soldiers and leaders to maneuver while firing service ammunition or SRTA. An LFX is most productive when preceded by non-live-fire practice (dry fire or dry run). However, dry fire should *always* precede live fire, with either TES or SESAM. An AAR should follow each repetition, and units should only proceed to the next level of difficulty after mastering the current step. The training in an LFX occurs in the following sequence, which can be shortened if time and ammunition dictate.

1. Dry fire to practice unit drills and individual tasks.
2. Conduct an AAR.
3. Practice drills, tasks a second time with limited use of ammunition to show the complex coordination of fire and maneuver required.
4. Conduct an AAR.
5. Practice exercise a third time with full use of ammunition to enhance realism and build confidence.
6. Conduct an AAR.

SITUATIONAL TRAINING EXERCISE

A-15. STXs are mission-oriented, limited exercises used to train a unit to perform a single collective task or a series of related tasks and drills. With their distinct start and stop points, STXs represent a segment of battle. An STX is more flexible and complex than a single T&EO or drill. In fact, an STX typically includes drills, leader tasks, and separate individual tasks, and involves a unit's "slice" of external combat assets. An STX often precedes an FTX or CFX, and can be conducted in the CACTF by creating a brief situation and mission. For example, the MP company commanders might want to teach his Soldiers about straggler control. They can string the tasks together into small scenarios such as STXs, describe a tactical situation that produces stragglers from nearby units, use other Soldiers to represent the stragglers, and train Soldiers to process stragglers at straggler collection and transport points.

FIELD TRAINING EXERCISE

A-16. FTXs allow units to execute collective tasks under conditions that closely resemble actual urban terrain. FTXs are suitable for the CACTF.

- An FTX requires significant time, resources, and planning. The time is well spent if units master their collective tasks. An FTX is an excellent way to conduct the *run* stage of collective training. Additionally, OPFOR and MILES greatly enhance the training value of the FTX.
- A unit's urban operations FTX training plan addresses ROE, maneuver restrictions, and the controller plan. Soldiers must know the exercise's training objectives before they begin, so they can focus on the correct tasks. Every participant, regardless of position or rank, must fully understand all safety requirements.

- To make the most of their training time in the field, leaders must plan in great detail. The concept of multiechelon training demands that they consider what training can occur at each level in each event.

EVALUATORS

A-17. Soldiers from the training unit or from another unit can evaluate urban operations training, especially during the *run* stage of training in an STX, FTX, or CFX. A dedicated evaluator provides objective and thorough comments on the unit's performance throughout all stages of the exercise. Evaluators monitor the critical point of the action but must avoid interfering. They must not disrupt unit performance or undermine realism. They must be trained and given standardized evaluation guides if their feedback is to be meaningful; MTPs and T&EOs are a good start. Evaluators facilitate AARs alone or along with the unit leader.

TACTICAL ENGAGEMENT SYSTEM

A-18. The TES realistically replicates the effects of actual fire in training. It uses laser adapters on the weapons and laser-sensitive equipment on the Soldiers and vehicles. A TES weapon emits a laser beam when fired. Accurate fire strikes the laser-sensitive equipment and scores a hit. The TES requires Soldiers to aim accurately and to use cover and concealment effectively. When used with an OPFOR, TES offers realistic training. It is an excellent training aid for an FTX or STX. It is also effective at the shoot house as a substitute for live fire. The two major shortcomings of TES are that it allows Soldiers to seek cover behind structures that normally would not protect them, and that it does not replicate suppression.

NAVIGATION

A-19. Navigational techniques for urban areas differ from those needed on other terrain.

- A typical 1:50,000-scale military map provides insufficient detail for accurate navigation. Therefore, Soldiers must learn to navigate using commercial maps, sketches, and aerial photographs. Leaders must learn how to plan operations using these items.
- Underground navigation is a special challenge. Soldiers must know subway plans and be able to identify tunnels, elevated tracks, stairways, and other critical features. They must learn how to measure distance, remain oriented, and mark their routes.
- Destruction often erases urban features that could aid in navigation. Soldiers must learn to orient using distinctive, difficult-to-erase features such as bridges, waterways, towers, and intersections.
- Leaders conduct navigational exercises in an actual urban area. Soldiers must move from point to point while controllers at each point monitor and verify Soldiers' progress.
- The GPS is a navigational aid that triangulates with satellites to calculate position. However, this system may be affected by nearby tall buildings, and its effectiveness decreases inside buildings or underground.

LIMITED VISIBILITY TRAINING

A-20. Urban operations involve day and night operations. FTXs, STXs, CFXs, and CPXs are the most valuable exercises for night training.

- Field-expedient lighting, NVDs, illumination, and electrical lighting must be exploited when the tactical situation allows. Soldiers must learn discipline in their use of light, and also learn the lighting techniques that allow them to carry out their urban training program.

- Situational awareness is the Soldier's knowledge of his location, the locations of friendly forces, and the locations of the enemy on the battlefield. Situational awareness is the key to avoiding fratricide, especially during limited visibility operations. Units must plan to use all the equipment and devices available to mark friendly units and to avoid fratricide, without exposing themselves to enemy observation.
- Reverse-cycle training should be part of an urban training program. Limited visibility fighting skills are perishable, but they can be sustained and improved by effective training. The goal is for every Soldier to operate as effectively at night as during the day.

SAFETY AND AMMUNITION CONSIDERATIONS

A-21. When using ammunition and pyrotechnics, all associated safety requirements must be followed. This paragraph describes some of the safety equipment and training procedures to consider:

SPECIAL EFFECT SMALL-ARMS MARKING SYSTEM

A-22. SESAMS is 5.56-mm and 9-mm color-marking training ammunition. It is designed for close combat training. It is used with the close combat mission capability kit (CCMCK) adapter kit, which is installed on the M4/16 rifle, M9/M11 pistol, and M249 SAW. Different colors can be assigned to OPFORs for engagement identification and fratricide tracking. SESAMS has a 30-meter effective range and activates TES (MILES), so the two are used together. Anytime during a SESAMS exercise that a Soldier loses any of his protective gear, he must immediately cover his face and eyes with his hands. He must stay put until the engagement ends, or until he is told that it is safe to replace his safety equipment. Everyone, to include observer-controllers, must wear the protection shown in Table A-2.

Table A-2. SESAMS protection.

SKIN	Wear gloves, and cover all exposed skin. At close ranges, the SESAMS projectile can break skin.
HEAD	Take every precaution to protect your head.
EYES	Keep your eyes completely enclosed within eye protection. You can use standard Army sun, wind, and dust goggles.
EARS	Wear hearing protection to prevent hearing loss.
MOUTH	Keep your mouth covered.

DANGER

Failure to wear the required protection during SESAMS exercises can result in injury or death.

M84 STUN GRENADE

A-23. The hand-deployed M84 stun grenade serves as a reliable, effective, nonlethal means of neutralizing, diverting, and disorienting enemy personnel. It is used tactically as well as in training.

A-24. The M84 produces an intense (1 to 2.5-million candlepower) flash and, at 1.5 meters, a 170 dBs bang. Even though this is too low to produce permanent hearing damage, one-time use Soldiers are required to wear hearing protection anytime noise levels exceed 85 dBs.

A-25. The M84 stun grenade has an M201A1 fuze and a pyrotechnic output charge in a synthetic casing with a one- to two-second delay fuze, which cannot be cooked off. The grenade body has a nonfragmenting metal hexagonal casing, which, once deployed, resists rolling.

SMOKE GRENADES

A-26. Smoke grenades are used in urban combat to cover movement, deceive enemy forces as to the exact location of the attacking force, and for signaling. Smoke grenades should not be used in buildings and subterranean passages as they displace oxygen and protective masks do not filter the smoke or provide oxygen. Buildings in urban areas channel air currents, which influence emplacement of smoke grenades. One grenade should be used to test the airflow before employing additional grenades.

WARNING

Avoid using smoke grenades in buildings and in subterranean passages.

OPPOSING FORCES

A-27. OPFOR should be used as often as possible to enhance the realism of training, especially for FTXs, CFXs, and STXs. Thebe realistic, the OPFOR must have defined countertasks. These countertasks must support "Blue Force" training, and should prevent accidental contact between the Blue Force and the OPFOR. Leaders should develop a thorough control plan for OPFOR scenarios. This plan should describe uniform and vehicle markings, ammunition, safety instructions, controller guidance, and guidelines for handling enemy prisoners of war (EPW) and noncombatants. The ROE must clearly specify the limits of close combat between Soldiers.

- OPFOR heighten Soldiers' interest by giving them a real opponent. This fosters unit competition and realism, especially when the OPFOR is used in addition to TES and SESAM.
- OPFOR challenges leaders' ability to improvise. However, this free-play scenario should be monitored to ensure the unit meets the training objectives.
- OPFOR can use threat urban operations and doctrine to determine their course of action. If available, OPFOR uniforms, mock documents, overlays, and equipment add realism and intelligence to the scenario. Leaders conduct preliminary training for the OPFOR, which helps the OPFOR recreate the threat's urban tactics.
- Noncombatants should be integrated into training scenarios.

PRECISION MARKSMANSHIP TRAINING

A-28. Precision marksmanship training should be conducted before training at the UAC, CACTF, and the Shoot House, which accommodate training in target discrimination (FM 3-22.9). This is the best use of the unit's time and resources.

PHYSICAL READINESS TRAINING

A-29. Combat in urban areas places extraordinary physical demands on Soldiers. Therefore, physical readiness training, with emphasis on upper body development, is paramount to any urban operations training program.

VARIED SCENARIOS

A-30. Leaders must vary training scenarios to prevent redundant and predictable tactical behavior. Using predictable avenues of approach into the facility produces stereotypical tactical solutions.

BUILDING FACADE WALLS, WINDOWS, AND DOORS

A-31. The use of façade walls, windows and doors, in conjunction with SESAMS in the interior of some stations at the UAC, provides for more realistic training. This method of training reinforces the fact that walls, windows, and doors in most cases *do not* stop the penetration of service ammunition. Plastic wrap sheeting, tar paper, or some other type of material can be used to allow SESAMS penetration (Figure A-1).

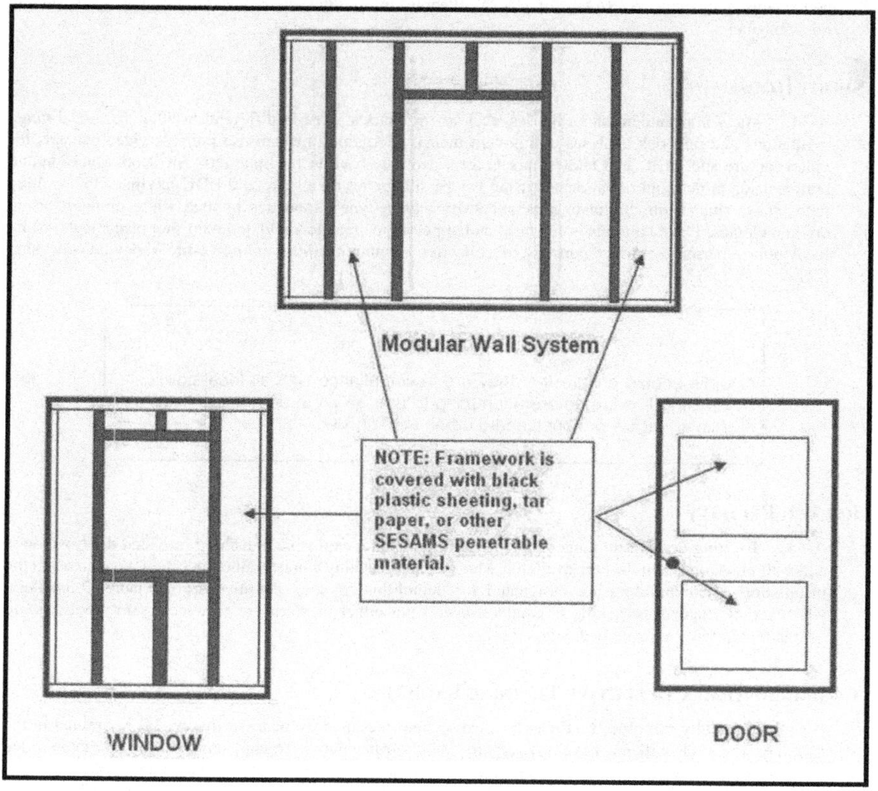

Figure A-1. Shoot-through walls and doors.

USE OF OLDER OR NONSTANDARD FACILITIES

A-32. Until installations receive funding for the UAC, SH, BF, and CACTF, they can use the older MOUT Assault Course (MAC) and Collective Training Facility (CTF) as described in the 2002 edition of TC 90-1, which this edition supersedes. Installations may also use nonstandard shoot houses and other nonstandard urban facilities. These facilities may be adapted as follows:

URBAN ASSAULT COURSE

A-33. Generally, the MAC can be used to train most of the UAC tasks. Of course, the MAC lacks targetry and instrumentation. Instead, units can record training with portable video cameras. During the AARs, they can play back the videos on VCRs hooked to TVs or on TV/VCR combos. OPFOR combatants will have SESAMS or MILES. People in civilian clothes can play the parts of the noncombatants.

SHOOT HOUSE

A-34. Many nonstandard shoot houses, such as tire houses, exist in different installations. Local range regulations and unit risk analyses will govern their use. Again, units can use portable video cameras for video capture and VCRs and television sets for video play back to facilitate an AAR. Both precision and nonprecision targets can be fabricated using E-type silhouettes on a stake, old BDUs/civilian clothes filled with straw, and so on. Simulate noncombatants with E-type silhouettes painted white or dummies in civilian clothes. Place balloons in the head and upper torso areas to simulate lethal shot placement. Where local policy permits, convert portions of collective training facilities into live-fire shoot houses using SRTA.

> **CAUTION**
>
> Units should ensure that they are in compliance with all local policies affecting live fire before attempting to use SRTA in an older collective training facility or nonstandard urban training site.

BREACH FACILITY

A-35. Existing demolition ranges may be used to conduct explosive breaching provided door, window, and wall units, or facsimiles, are available. Mechanical breaching can be trained in cantonment areas on the installation where building are designated for demolition, if such facilities are available. Regardless, whenever mechanical breacking is conducted, unit personnel must ensure that local safety policies are available, briefed, and complied with.

COMBINED ARMS COLLECTIVE TRAINING FACILITY

A-36. Generally the older CTF can be used to train the majority of tasks that would be trained in the CACTF. The CTF will not have any targetry and instrumentation. Again, units can use portable video cameras for video capture and VCRs and television sets for video play back to facilitate an AAR. Targets can be OPFOR with MILES. Role players in civilian clothes can be introduced to play the part of noncombatants. If armored vehicles are used, units should ensure that local range policies allow for their use. Older CTFs are normally either a 16 or 32 building design. Appropriate unit-level training, platoon, company, or battalion, should be planned based on the size of the facility.

Appendix B

Unit Training Matrix

Unit training plans are prepared IAW FM 7-1, which states that "Planning links the unit METL and the execution of battle-focused training.". The commander begins his planning with guidance from a higher echelon and an in-depth assessment of METL tasks (in this case, mission-essential urban tasks). Based on his own observations, feedback from his subordinates, and other sources the commander determines which tasks should be trained, which should be trained first, and how much time should be allotted for each. Tables B-1 through B-5 list training tasks by the WFF to be trained at each facility.

Table B-1. Army unit training matrix, brigade tasks.

Brigade Tasks	*SHOOT HOUSE* *Semiannual*	*BREACH HOUSE* *Semiannual*	*UAC* *Quarterly*	*CA/CTF* *Semiannual*	*CTC* **As Scheduled*	*SIMULATION* *Annual*
Develop Intelligence						
Perform Intelligence Operations					X	X
Perform S-2 Operations					X	X
Deploy/Conduct Maneuver						
Attack					X	X
Perform Air Assault					X	X
Defend					X	X
Lodgment					X	X
Exercise Command and Control						
Command and Control the Brigade					X	X
Maintain Communications					X	X
Coordinate Air Defense Operation					X	X
Conduct Noncombatant Evacuation					X	X
Develop a Media Plan					X	
Conduct Mediation and Negotiation					X	
Employ Firepower						
Execute Fires					X	X
Employ Air Support					X	X
Employ Attack Helicopters					X	X
Synchronize Fire Support					X	X
Synchronize CAS					X	X
Conduct Targeting Process					X	X
Protect the Force						
Perform Mobility/Survivability Operations					X	X
Provide Engineer Support					X	X
Coordinate CBRN Operations					X	X
Conduct Operations Security					X	X
Perform Sustainment						
Evacuate Casualties					X	X
Treat Casualties					X	X
Sustain the Force					X	X
Conduct a Civil-Military Operation					X	X
Provide Humanitarian Support					X	X

Table B-2. Army unit training matrix, battalion tasks.

Battalion Tasks	SHOOT HOUSE Semiannual	BREACH HOUSE Semiannual	UAC Quarterly	CA/CTF Semiannual	CTC As Scheduled	SIMULATION Annual
Develop Intelligence						
Conduct S-2 Operations				X	X	X
Deploy/Conduct Maneuver						
Attack a Built-Up Area				X	X	X
Plan for Urban Operation				X	X	X
Defend an Urban Area				X	X	X
Conduct an Infiltration/Exfiltration				X	X	X
Perform Cordon and Search Operations in an Urban Area				X	X	X
Plan for Urban Operations				X	X	X
Conduct Tactical Movement				X	X	X
Conduct Presence Operation in a Stability Environment				X	X	X
Defend an Urban Area				X	X	X
Traverse a Built-up Area				X	X	X
Conduct a Mobile Defense				X	X	X
Employ Firepower						
Integrate Fires and Effects Support				X	X	X
Establish the Fire Support Element				X	X	X
Protect the Force						
Conduct Mobility or Survivability Operations				X	X	X
Secure Civilians During Operation				X	X	X
Conduct Passive Air Defense Measures				X	X	X
Conduct Active Air Defense Measures				X	X	X
Employ Operations Security Measures				X	X	X
Integrate Air Defense Support				X	X	X
Perform Sustainment						
Conduct S-1 Operations				X	X	X
Conduct S-4 Operations				X	X	X
Conduct a Civil-Military Operation				X	X	X
Exercise Command and Control						
Conduct S-6 Operations				X	X	X
Conduct Noncombatant Evacuation				X	X	X
Conduct S-3 Operations				X	X	X
Conduct Mediation and Negotiations				X	X	X
Plan Roadblocks or Checkpoints				X	X	X
Establish Liaison				X	X	X

Table B-3. Army unit training matrix, company tasks.

Company Tasks	*SHOOT HOUSE* Semiannual	*BREACH HOUSE* Semiannual	*UAC* Quarterly	*CA/CTF* Semiannual
Develop Intelligence				
Conduct an Area or Zone Reconnaissance				X
Reconnoiter a Built-Up Area				X
Deploy/Conduct Maneuver				
Conduct Cordon and Search				
in a Built-Up Area				X
Attack a Built-Up Area				X
Conduct Presence Operations				X
Conduct Tactical Movement				
in a Built-Up Area				X
Defend in Urban Environment				X
Conduct an Infiltration (Co/Plt)				X
Conduct Operations with Armored or				
Mechanized Infantry Vehicles				
in an Urban Environment				X
Conduct Civil Disturbance Operations				X
Establish Checkpoints				X
Employ Firepower				
Integrate Direct Fires				X
Integrate Indirect Fires				X
Protect the Force				
Maintain Operations Security				X
Conduct Passive Air Defense Measures				X
Conduct Passive Air Defense Measures				X
Perform Sustainment				
Treat Casualties				X
Evacuate Casualties				X
Secure Civilians during Operations				X
Conduct Resupply Operations				
Process Captured Documents and				X
Equipment				X
Handle Enemy Prisoners of War				
Exercise Command and Control				
Prepare for Combat				X
Establish Communications				X
Conduct Consolidation and Reorganization				X
Conduct Negotiation				X
Conduct Troop-Leading Procedures				X

Table B-4. Army unit training matrix, platoon/squad tasks.

Platoon/Squad Tasks	*SHOOT HOUSE Semiannual*	*BREACH HOUSE Semiannual*	*UAC Quarterly*	*CA/CTF Semiannual*
Develop Intelligence				
Report Tactical Information	X		X	X
Conduct an Area or Zone Reconnaissance	X		X	X
Reconnoiter a Built-Up Area	X		X	X
Deploy/Conduct Maneuver				
Conduct Tactical Movement in a Built-Up Area	X		X	X
Assault a Building	X		X	X
Conduct a Strongpoint Defense of a Building				X
Conduct an Infiltration or Exfiltration				X
Conduct Operations with Armored or Mechanized Vehicles in an Urban Environment	X			X
Conduct a Presence Patrol	X			X
React to a Civil Disturbance				X
Establish a Checkpoint				X
Search a Building	X		X	X
Employ Firepower				
Employ Fire Support				X
Protect the Force				
Maintain Operations Security				X
Conduct Passive Air Defense Measures				X
Conduct Active Air Defense Measures				X
Perform Sustainment				
Secure Civilians During Operations	X		X	X
Perform Resupply Operations				X
Treat Casualties	X		X	X
Evacuate Casualties	X		X	X
Handle Enemy Prisoners of War				
Exercise Command and Control				
Conduct Troop-Leading Procedures	X		X	X
Establish Radio Communication	X		X	X
Conduct Consolidation and Reorganization	X		X	X
Prepare for Combat	X		X	X

Table B-5. Army unit training matrix, individual tasks/techniques.

Individual Tasks/Techniques	*SHOOT HOUSE Semiannual*	*BREACH HOUSE Semiannual*	*UAC Quarterly*	*CA/CTF Semiannual*
Skill Level 1				
Perform Movement Techniques in an Urban Environment, 071-326-0541	X	X	X	X
Prepare Positions for Individual and Crew-Served Weapons in an Urban Environment, 071-326-0550	X	X	X	X
Select Hasty Fighting Positions in an Urban Environment, 071-326-0557	X	X	X	X
Employ Hand Grenades, 071-325-4407	X	X	X	X
Engage Targets with an M16-Series or M16A2 Rifle, 071-311-2007	X	X	X	X
Execute a Mechanical Breach, 071-440-0027			X	X
Skill Level 2				
Conduct an Explosive Breach	X	X	X	X
Conduct a Breach, 071-440-0029	X	X	X	X
Prime Explosive Nonelectrically, 051-193-1003	X	X	X	X
Construct a Nonelectric Initiating Assembly with Modernized Demolition	X	X	X	X

Appendix C

Stability and Support Operations Unit Training Matrix

When assigned a stability or support mission, a well-trained unit must be able to rapidly shift its focus from war fighting to stability and support and also from stability and support to war fighting. During a stability or support operation, the unit performs numerous activities. Essentially, the unit accomplishes these activities through execution of tactical missions and tasks. While stability and support operations can occur anywhere, it is most probable that they will occur in an urban environment. Tables C-1 through C-4 list training tasks by the WFF to be trained at each facility.

Table C-1. Stability and support operations unit training matrix, brigade tasks.

Brigade Tasks	*SHOOT HOUSE*	*BREACH HOUSE*	*UAC*	*CA/CTF*	*CTC*	*SIMULATION*
Develop Intelligence						
Collect and Disseminate Information					X	X
Deploy/Conduct Maneuver						
Demonstrate a Show of Force					X	
Employ Psychological Operation					X	X
Employ Firepower						
Provide Fire Support					X	X
Protect the Force						
Protect the Force					X	
Perform Sustainment						
Restore Law and Order					X	
Exercise Command and Control						
Establish an Evacuation Control Center/Conduct Evacuation Control Center Operations					X	
Negotiate					X	
Perform Noncombatant Evacuation Operations					X	
Plan a Media Visit					X	
Develop Intelligence						
Collect and Disseminate Information				X	X	X

Table C-2. Stability and support operations unit training matrix, battalion tasks.

Battalion Tasks	*SHOOT HOUSE*	*BREACH HOUSE*	*UAC*	*CA/CTF*	*CTC*	*SIMULATION*
Deploy/Conduct Maneuver						
Interdict Smuggling Operations				X	X	
Disarm Belligerents				X	X	
Handle Captured Insurgents or Belligerents				X	X	
Employ Psychological Operation				X	X	
React to Civil Disturbance				X	X	
Employ Firepower						
Provide Fire Support				X	X	X
Protect the Force						
Protect the Force				X	X	
Restore Law and Order						
Identify and Process Detainees				X	X	
Deliver Supplies or Humanitarian Aid				X	X	
Control Civilian Movement				X	X	
Prepare Traffic Control Plan				X	X	
Exercise Command and Control						
Negotiate				X	X	
Perform Noncombatant Evacuation Operations				X	X	
Conduct Marshalling Force Operations				X	X	
Negotiate a Belligerent Force Checkpoint				X	X	
Plan a Media Visit				X	X	
Perform a Cordon and Search				X	X	
Coordinate and Monitor Convoy Security				X	X	

Table C-3. Stability and support operations unit training matrix, company tasks.

Company Tasks	*SHOOT HOUSE*	*BREACH HOUSE*	*UAC*	*CA/CTF*	*CTC*	*SIMULATION*
Develop Intelligence						
Collect and Disseminate Information				X	X	
Deploy/Conduct Maneuver						
Interdict Smuggling Operations				X	X	
Disarm Belligerents				X	X	
Handle Captured Insurgents or Belligerents				X	X	
Defend Convoy				X	X	
Apprehend/Detain Noncombatants				X	X	
Perform Cordon and Search				X	X	
Employ Firepower						
Provide Fire Support				X	X	
Protect the Force						
Establish a Checkpoint				X	X	
Operate a Checkpoint				X	X	
Protect the Force				X	X	
Perform Sustainment						
Process Captured Documents and Equipment				X	X	
Exercise Command and Control						
Negotiate a Belligerent Force Checkpoint				X	X	
Conduct Marshaling Force Operations				X	X	
Negotiate a Belligerent Force Checkpoint				X	X	
Plan a Media Visit				X	X	
Coordinate and Monitor Convoy Security				X	X	
Link Up with a Convoy				X	X	

Table C-4. Stability and support operations unit training matrix, platoon/squad tasks.

Platoon/Squad Tasks	*SHOOT HOUSE*	*BREACH HOUSE*	*UAC*	*CA/CTF*	*CTC*	*SIMULATION*
Deploy/Conduct Maneuver						
Disarm Belligerents				X	X	
Handle Captured Insurgents or Belligerents				X	X	
Apprehend/Detain Noncombatants				X	X	
Search a Building	X			X	X	
React to a Sniper			X	X	X	
Conduct Platoon Riot Control Formation				X	X	
Protect the Force						
Establish a Checkpoint				X	X	
Operate a Checkpoint				X	X	
Protect the Force				X	X	
Exercise Command and Control						
Negotiate a Belligerent Force Checkpoint				X	X	
Exercise Command and Control						
React to a Media Interview				X	X	

Appendix D

Leader Development Training

To be successful in combat, the Army must continually train to develop and maintain combat-ready Soldiers, leaders, and units that can perform assigned tasks to specific standards.

SIMULATIONS TRAINING

D-1. Commanders train their staffs to plan, coordinate, and synchronize their units' operations from initial mobilization and deployment. These operations include the full spectrum of Army operations; offense, defense, stability, and support. Current simulations lack adequate replication of urban scenarios, terrain, and weapons effects. The introduction of WARSIM 2000 and OneSAF should help to solve these simulations modeling deficiencies. For lower echelon units, battalion and below, constructive simulations complement collective battle staff training that occurs during virtual simulation similar to the close combat tactical trainer and live events. For echelons above battalion, it becomes difficult to use simulators or live maneuver to train. Studies to date indicate the crossover point for simulation versus simulator is at the battalion level. Some training events from battalion and below can be trained effectively using various types of simulators. Above battalion level, effective training almost has to be conducted with a simulation (Figure D-1).

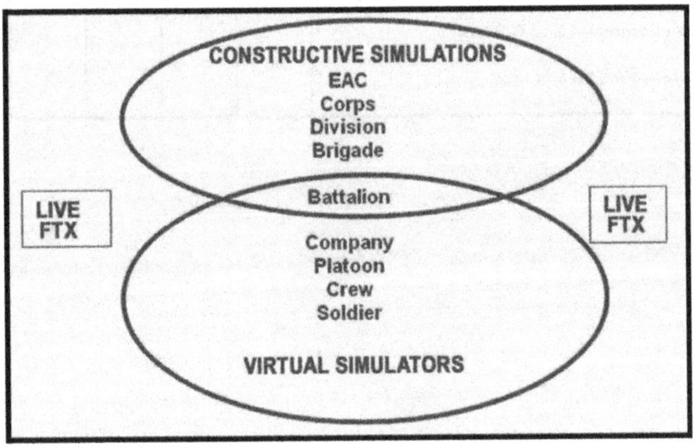

Figure D-1. Echelons and uses.

D-2. Simulations that support leader and battle staff training focus on collective tasks rather than individual leader tasks. Simulations, used in this context, rely on the human dynamics in the command setting and use real-time actions in a tactical environment with a unit's go-to-war systems. This process forces commander and staff interaction that provides feedback and helps refine the integration process. Battle staff training encourages team building, task coordination, and standardization of staff actions.

- Constructive simulations can provide commanders large areas for conducting training, access to sensitive areas through computer and mapping technology, and futuristically provide a much better level of fidelity in terms of urban operations terrain and scenario replication.
- Simulations can portray large, capable, and doctrinally correct OPFORs. All of the Soldiers portrayed in the units in a simulation do not have to be present in order for the training to be realistic. An attractive feature of simulations has always been the reduced staffing necessary to conduct this type of training exercise.
- Simulations can stress CPs and provide the realistic conditions under which commanders must make decisions. Brigade and below simulation (BBS) and corp battle simulation (CBS) are well known for the stressful environments created during an exercise. The stress is realistic and can assist in conditioning the training audience before CTC rotations.
- Simulations allow different units to train under the same conditions and to the same standards. This technique is a unique feature of simulations as the exact weather, opposing force organization, location of units, and equipment can be used to train many different units on the same terrain with the same conditions.
- The size and number of personnel in a battle staff vary depending on the echelon and other demands unique to that commander. However, the WFF are represented by the following staff officers: intelligence, operations, fire support, communications, logistics, air defense, and engineer.
- Characteristics and benefits of simulations used to support battle staff training follow:
  -- Build and sustain staff teamwork.
  -- Build and sustain effective and efficient organizations.
  -- Provide training and validation of command and staff procedures (SOPs, TACSOPs, FSOPs).
  -- Provide proficiency to the individual staff officer, while building proficiency in a fully integrated staff.
  -- Allow practice of battle drills and tactical missions (movement to contact, hasty defense).
  -- Allow commanders to determine if their staff and subordinate commanders understand the commander's intent and concept of operations.

D-3. The major development efforts for FAMSIM is WARSIM 2000. WARSIM 2000, in conjunction with OneSAF, exploits new technologies to enable commanders and staffs at battalion through EAC to train in a realistic simulation environment (Table D-1). The increased realism of WARSIM 2000 over existing models allows units to use Army Battle Command Systems (ABCS) to synchronize across the complete range of operations. WARSIM 2000's design allows warfighting CPs to use organic ABCS equipment to interact with the simulation from training sites in the field, as well as from simulation centers. WARSIM 2000 will also be capable of showing a joint and combined environment across the operational continuum.

Table D-1. Urban operations simulations crosswalk.

CAMTF Training Tasks	*WARSIM*	*OneSAF*
Corps	X	
Division	X	
Brigade		X
Battalion		X
Company		X
Platoon		X

D-4. Future simulation capabilities simultaneously integrate live, virtual, and constructive environments. Constructive simulation is a viable solution for battalion and above training and the federation of WARSIM and OneSAF models should alleviate the following training simulation problems:

- Eliminate the need for swivel chair interface.
- Allows focus on subordinate units fighting the fight.
- Digital and analog feeds to C4I systems including FBCB2.
- May provide for distance learning opportunities.
- Facilitates simultaneous multi-echelon urban operations training

D-5. At company and below, the best training solution is live and virtual. This is accomplished by linking the Close Combat Tactical Trainer (CCTT) to OneSAF and VLET to Engagement Skills Trainer (EST) technology. Entity-based models like OneSAF can allow the leader to fight from a TOC or vehicle, if linked to virtual or live environments. It also has the potential for distance learning.

PHASES OF TRAINING

D-6. The Army Urban Operations Training Strategy consists of initial, intermediate, and culminating stages. These phases of training, shown in Figure D-2, describe the sequence in which urban operations training tasks should be conducted.

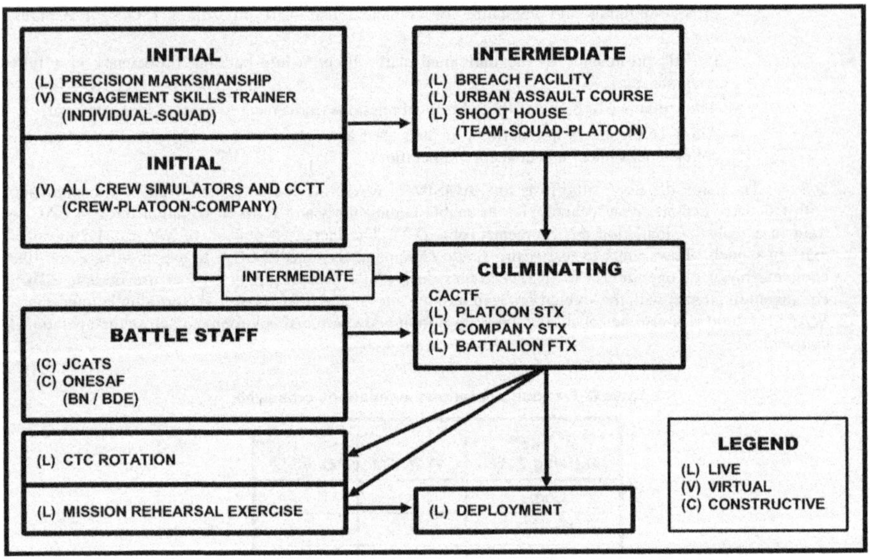

Figure D-2. Phases in the Live, Virtual, and Constructive Training Strategy.

Initial Phase

D-7. The initial phase of training covers those individual and collective tasks that Soldiers and leaders should be proficient at before arriving at any of the urban operations training sites. They consist primarily of basic marksmanship skills and Soldier tasks needed to perform any tactical operation.

Intermediate Phase

D-8. The intermediate phase consists of training conducted at the breach facility, shoot house, and the urban assault course. Urban operations training tasks should be conducted here before advancing to the Combined Arms Collective Training Facility with the exception of Aviation, Armor, and other branches that have no practical use for the Shoot House, Breach Facility, or Urban Assault Course.

Culminating Phase

D-9. The combined arms collective training facility is the culminating phase of urban operations training and focuses on multiechelon collective skills for all branches of the combined arms team. Urban operations leader training currently exists in our leader professional development courses. Units have developed internal urban warfare training programs and TRADOC schools have assembled technical mobile training teams (MTTs) for marksmanship, combatives, and breaching techniques.

Education Phase

- Urban warfare reading/video list.
- Urban operations equipment.
- Review of doctrinal publications (FM 3-21.10, FM 3-06.11, for Infantry) and corresponding publications for other branches.
- Small-unit urban movement techniques and considerations.

Planning and Execution Phase

- Prepare company OPORD or WFF-related products.
- Brief platoon/company OPORD/WFF-related products to commander or staff representative.
- Participate as briefer in urban environment TEWT.

Sustainment Phase

- Participate in company/platoon TEWT or other urban operations event.
- Execute/facilitate a company or platoon TEWT.

EXAMPLE LEADER'S COURSE POI

D-10. An example urban leaders' course POI, which is intended to provide unit trainers with a tailorable urban operations training program, is shown in Table D-2. It is a *train-the-trainer* course for team leaders through platoon leaders to train tactics, techniques, and procedures of advanced urban operations skills. The length and focus of the POI can be tailored to support the commander's training objectives and unit METL. The target audience is newly arrived, or soon to be, small unit leaders. The recommended course length is about two weeks or 120 to 140 hours of instruction.

REALISTIC TRAINING OFF FEDERAL FACILITIES

D-11. Realistic training for urban operations is critical to force readiness. Training opportunities that are both realistic and challenging may be unavailable at federal installations, or at facilities specifically designed for urban operations training. Training in off-post civilian urban settings may become necessary in order to support readiness and training requirements.

D-12. The Department of Defense has developed the following SOP for planning and approving such events. The SOP assists in balancing the needs of the military training and of the civilian community/ies to be used for realistic urban training events.

Table D-2. POI for urban leader's train-the-trainer course.

Course Overview	• Introduction	
Precision Marksmanship	• Firing techniques. • Target discrimination.	
Breaching Techniques	• Explosive. • Ballistic. • Mechanical.	

Table D-2. POI for urban leader's train-the-trainer course (continued).

Offensive Operations	• Planning considerations	-- METT-TC. -- Snipers. -- CASEVAC. -- Communications. -- ROE.
	• Platoon attack a built-up area (task organization)......................................	-- Assault (isolate, enter, clear). -- Rooms, hallway, stairs, superstructure, subsurface. -- Markings. -- Movement techniques. -- Streets. -- Alley ways. -- Obstacles. -- Other danger areas. -- Open areas. -- Roof tops.
	• Combined arms	-- Armor/mechanized platforms. -- Aviation assets. -- Artillery. -- Engineer.
	• Limited visibility	
Defensive Operations	• Planning considerations	-- METT-TC. -- Snipers. -- CASEVAC. -- Communications. -- ROE.
	• Characteristics of urban areas	
	• Hasty defense	-- Occupation and preparation of positions. -- Hasty firing positions. -- Improving the defense.
	• Deliberate defense, including priorities of work and other considerations	
	• Defensive plan at platoon level.	-- Defense of a strongpoint. -- Defense against armor.
	• Combined arms	-- Armor/mechanized. -- Aviation assets. -- Artillery. -- Engineer.
	• Limited visibility • Fundamental combat skills	

Table D-2. POI for urban leader's train-the-trainer course (continued).

Stability and Support Operations	• Principles of stability and support.	
	• Stability and support activities	-- NEO. -- Humanitarian assistance. -- Disaster relief. -- Show of force. -- Attack and raids.
	• Rules of engagement application • Convoy operations • Checkpoints • Fire support • Sniper employment/counter sniper measures • Civil disturbance	
Urban Training Facilities	• Breach facility • Urban assault course • Shoot house • CACTF	
New Technologies	• Rifle launch entry munition	
	• Unmanned vehicles	-- UAV. -- UGV. -- Ladders. -- Night vision devices.

STANDING OPERATING PROCEDURES FOR PLANNING AND APPROVING REALISTIC URBAN TRAINING

1. **REFERENCES.**

    10 USC § 167, *Unified Combatant Command for Special Operations Forces.*

    10 USC §§ 3013, 5013, 8013.

    Department of Defense Directive 5111.10.

    Unified Command Plan 99.

2. **PURPOSE**. This SOP establishes uniform planning and approval procedures for realistic urban training (RUT) events.

3. **APPLICABILITY**. This SOP applies to all RUT events in civilian urban settings within the United States, its territories, and its possessions; and to Active or Reserve Forces, including National Guard Forces in Federal service. This policy does not apply to National Guard Forces under in-state status (Title 32), or to aviation operations that do not require an exemption from standard Federal Aviation Regulations. The training event should not be located on federal state, local, or private installations or facilities.

4. **DEFINITIONS**. For the purposes of this SOP, the following definitions apply:

a. **Urban Operations. This term** includes (see FM 3-06) all military actions planned and conducted on a topographical complex, and on adjacent natural terrain, where man-made construction is the dominant feature. It includes combat in towns and cities, including house-to-house and street-by-street fighting.

b. **Realistic Urban Training.** RUT is high-intensity, close-quarter battle training using live or nonlethal fires; demolitions or explosives such as breaching charges; and air and naval supporting platforms at the objective, in ***civilian urban settings***.

c. **Civilian Urban Setting.** A civilian urban setting is any area(s) located in or near a civilian community. Civilian urban settings will not be developed or established for training events, especially where training activities would affect civilian residences or commercial areas due to noise, traffic, and so on.

5. **PROCEDURES**. The following procedures apply to the planning and approval of RUT in civilian urban settings:

a. The combatant commander or the service ensures compliance with this SOP. Approval authority for RUT events may be delegated to a general or flag officer within the chain of command, or to the combatant common service conducting the training.

b. The approval authority reviews proposed training events to determine whether they are necessary and appropriate for unit readiness. Training must support unit METs and JMETs.

c. On approval of the proposed training, and at least 14 days before training starts, the combatant common service notifies the **SECR/PF** through the CJCS (by way of message) of the approved training. At a minimum, this notice includes the following:

- Name and location of the unit conducting training.
- Scope of training and activities planned such as live fire, breaches, or use of aircraft.
- Start and end dates.
- Training location.
- A statement that all necessary coordination has been accomplished IAW this SOP. The statement should include a list of the contacted official's names and their positions.

d. **Coordination with Civilian Officials**.

- While planning the training event, the commander coordinates with and obtains written approval to conduct the event from the appropriate federal, state, and local civilian authorities. Commanders will consult their legal and public affairs officers to identify the civilian officials, for example the mayor, borough chief, county commissioners or supervisors, with whom they must coordinate the training event. To do this, they consult local and Federal law enforcement agencies. At a minimum, they must consult a senior official with responsibility for each affected civilian urban setting.
- During coordination, the commander covers the details of the proposed training activities, such as whether the event involves live fire, rotary- or fixed-wing aircraft, or close quarters battle training; and whether the exercises will be conducted during the day, night, or both: and so on. It will also list actions to ensure public and participant safety, proposed public or resident notification and outreach procedures, and any other issues that might cause public interest or concern.
- As much as possible, the commander briefs civilian officials in person on the exercise concept.

- The US congressional representative or staff for the training district, and the US senators for the state, or their staffs, are advised of the proposed training event and given whatever information they request. Neither written approval nor concurrence from the representative or senators is required. However, if these officials raise objections or issues, and if those issues cannot be resolved, they are immediately elevated to the Joint Staff, J3, and the service or combatant commander point of contact.
- For each training site, license agreements with civilian or government property owners or managers are executed. Agreements address liability and conditions of acceptable usage of property. Property owners are informed of the nature of the training including written notification as to whether the activities will include use of breaching charges and, if so, how much; or live ammunition and, if so, which weapons and what ammunition. This notice will also state *if* rotary or fixed-wing aircraft will be used *and* whether the exercises will be conducted during the day, night, or both. Before the exercise can begin, all license agreements must be reviewed for legality. Copies of the license agreements will be provided to the senior civilian officials responsible for training area.
- A record of coordination activities is maintained for each training event. The record includes the names and positions of the officials *who* coordinate and approve the event. It also provides officials with a summary of the events such as date, time, and location of the training; types of activities; and planned public notification and outreach. A follow-on letter is distributed to the senior civilian officials responsible for the affected civilian urban settings. The letter confirms the training event *and* lists the names of the civilian officials who plan the training event.
- The unit may schedule recurring training events at a particular area or location by submitting a single MOA to the authorizing civilian officials. The terms of the MOA must comply with the requirements of this SOP. The service or the combatant commander reviews the MOA, which authorizes the recurring training, to ensure it complies with the procedures. Once the commander approves the MOA, the unit may conduct individual training events under it without further Joint Staff or OSD review. However, services and combatant commanders notify SO/LIC *and* the Joint Staff 14 days *before* the training in this MOA begins. Thereafter, services and combatant commanders must review these MOAs periodically to ensure *that* they continue to fulfil the intent.
- *If* nondisclosure agreements are required, command legal representatives must review them. Nondisclosure statements should clearly explain *that* the agreement does not preclude persons from informing others of general, nonsensitive information about the training exercise.
- Depending on OPSEC and safety considerations, commanders must adhere to the recommendations of civilian officials regarding community notification, outreach, and press procedures. Community notification and outreach procedures may include door-to-door notification by the local law enforcement agency. Commanders will advise civilian officials when they cannot follow their recommendations.

e. Other instructions include--

- Legal and PA representatives attend final planning meetings, and they remain on-site throughout all training activities.
- The on-site PAO handles all press queries after coordinating with the on-scene commander.
- All requested training must follow applicable statutes, law, DOD directives, other guidance, and implementing service directives.

Glossary

Section I. ACRONYMS AND ABBREVIATIONS

1SG first sergeant

A

AA assembly area
AAR after-action review/report
ABCS Army Battle Command System
ACU Army combat uniform
ADA air defense artillery
ANCOC advanced noncommissioned officer course
AO area of operation
AR Army Reserve
AT4 a shoulder-fired munition, formerly called a "light antiarmor weapon"
atk attack
ATWESS antitank weapon effect signature simulator
AVCATT armored vehicle combined arms tactical trainer

B

BBS brigade and below simulation
BCT brigade combat team
bde brigade
BDU (obsolete; *see* ACU)
BF Breach Facility
bn battalion
BNCOC basic noncommissioned officer course
BOLC Basic Officer Leadership Course
BOS (obsolete; *see* WFF)
BP battle position
BSA brigade support area
BSX battle simulation exercise

C

C2 command and control
CACTF Combined Arms Collective Training Facility
CASEVAC casualty evacuation
CBRN chemical, biological, radiological, nuclear
CBS corps battle simulation
CCMCK close combat mission capability kit
CCOC Captain's Career Officer Course
CCTT close combat tactical trainer
CCTV closed-circuit television
CFX command field exercise
CMTC Citizens Military Training Corps
co company
commo communications
CP command post
CPX command post exercise
CTC Combat Training Center
CTF collective training facility

D

DES demolition effects simulator
det cord detonation cord
div division
DLIC detachment left in contact
DS Digital Training Management System
DTMS direct support

E

EA engagement area
EAC echelons above corps
ENCATT engineer combined arms tactical trainer
engr engineer
EPW enemy prisoners of war
EST Engagement Skills Trainer

F

FAMSIM family of simulations
FO forward observer
FRAGO fragmentary order
FSCATT fire support combined arms tactical trainer
FSOP field standing operating procedures
FTX field training exercise

G

GREM grenade rifle entry munition

GSR ground surveillance radar

H

HC herachloroethane zinc

HEAT high-explosive antitank

I

IAW in accordance with

Inf Infantry

J

JANUS Joint Army-Navy Uniform Simulation

JCATS Joint Conflict and Tactical Simulation

JMET joint mission-essential task

JRTC Joint Readiness Training Center

JTF joint task force

JTX joint training exercise

L

LAW light antiarmor weapon

LFX live-fire exercise

LMG light machine gun

M

MAC MOUT assault course

MACOM major command

MAPEX map exercise

M/C/S mobility/countermobility/ survivability

MET mission-essential task

METL mission-essential task list

METT-TC mission, enemy, terrain, troops, time available, and civilians

MI military intelligence

MILES Multiple Integrated Laser-Engagement System

MOPP mission-oriented protective posture

MPAT multipurpose antitank (ammunition)

MPAT-OR multipurpose antitank-obstacle reducing

MRE meals ready to eat

MTT mobile training team

N

NBC (obsolete; *see* CBRN)

NEO noncombatant evacuation operations

NLT not later than

NTC National Training Center

O

OneSAF one semiautomated force (software used to create modeling and simulation programs with military applications)

op operation

OPFOR opposing force

OPORD operation order

OPSEC operations security

P

plt platoon

POI program of instruction

PPP power projection platform

PSP power support platform

R

ROE rules of engagement

ROI rules of interaction

RUT realistic urban training

S

S-1 personnel staff officer (adjutant)

S-2 intelligence staff officer

S-3 operations staff officer

S-4 logistics staff officer

S-5 civil affairs officer

S-6 command, control, communications and computer operations (C4 operations) officer

SAAF small-arms alignment fixture

SBCT Stryker brigade combat team

SDZ surface danger zone

SEP system enhancement program: a version of the M1A2 machine gun with enhanced digital command and control and second-generation forward-looking infrared

SESAMS	Special Effects Small Arms Marking System
SH	Shoot House
SOI	signal operation instructions
SOP	standing operating procedure
sqd	squad
SRTA	short-range training ammunition
STAFFEX	staff exercise
STX	situational training exercise

T

T&EO	training and evaluation outline
TACSOP	tactical standing operating procedure
TES	Tactical Engagement System
TEWT	tactical exercise without troops
TF	task force
TOW	tube-launched, optically tracked, wire-guided (missile)
TTP	tactics, techniques, and procedures

U

UAC	urban assault course
UO	urban operations

V

VCR	videocassette recorder
VLET	virtual leader effects trainer

W

WARSIM	warfighting simulation
WFF	warfighting function
WFX	warfighting exercise

X

XO	executive officer

Section II. TERMS

bustle rack	storage rack mounted on turrets of some armored vehicles
entity-based model	a representation of an actual or conceptual system that includes mathematical, logistical, or computer simulation
warfighting functions	group of tasks and systems (people, organization, information, and processes) united by a common purpose that commanders use to accomplish missions and training objectives. The warfighting functions are intelligence, movement and maneuver, fire support, protection, sustainment, and command and control. These warfighting functions replace the battlefield operating systems (BOSs).

References

SOURCES USED/DOCUMENTS NEEDED

These are the sources quoted or paraphrased in this publication. All of them are needed by and must be available to the intended user.

DOD DIRECTIVE

DOD Directive 2311.01E *DOD Law of War Program*. 9 May 2006.

FIELD MANUALS

FM 3-06	*Urban Operations*. 26 October 2006.
FM 3-06.11	*Combined Arms Operations in Urban Terrain*. 28 February 2002.
FM 3-20.12	*Tank Gunnery (Abrams)*. 15 August 2005.
FM 3-21.8	*The Infantry Platoon and Squad*. 28 March 2007.
FM 3-21.9	*SBCT Infantry Rifle Platoon and Squad*. 2 December 2002.
FM 3-21.10	*The Infantry Rifle Company*. 27 July 2006.
FM 3-21.11	*The SBCT Infantry Rifle Company*. 23 January 2003.
FM 3-21.20	*The Infantry Battalion*. 13 December 2006.
FM 3-21.21	*The Stryker Brigade Combat Team Infantry Battalion*. 8 April 2003.
FM 3-21.71	*Mechanized Infantry Platoon and Squad (Bradley)*. 20 August 2002.
FM 3-22.9	*Rifle Marksmanship M16A1, M16A2/3, M16A4, and M4 Carbine*. 24 April 2003.
FM 7-0	*Training the Force*. 22 October 2002.
FM 7-1	*Battle Focused Training*. 15 September 2003.
FM 25-4	*How to Conduct Training Exercises*. 10 September 1984.

SOLDIER TRAINING PUBLICATIONS

STP 21-1-SMCT	*Soldier's Manual of Common Tasks Skill Level 1*. 14 December 2007.
STP 21-24-SMCT	*Soldier's Manual of Common Tasks (SMCT) Warrior Leader Skill Levels 2-4*. 2 October 2006.

TRAINING CIRCULAR

TC 7-98-1	*Stability and Support Operations Training Support Package*. 5 June 1997.

Index

Made in the USA
Monee, IL
05 November 2026

67808350R00134